Disclaimer

The publisher of this book is by no way associated with the National Institute of Standards and Technology (NIST). The NIST did not publish this book. It was published by 50 page publications under the public domain license.

50 Page Publications.

Book Title: Characterization and Identification of Super-Effective Thermal Fire Extinguishing Agents. Final Report. NGP Project 4C/1/890

Book Author: William M. Pitts; Rodney A. Bryant; L G. Blevins; M L. Huber

Book Abstract: The use of halon 1301 for fire fighting is being phased out due to its deleterious effects on stratospheric ozone. This report summarizes the findings of a three-year study designed to characterize and identify super-effective thermal fire-fighting agents as possible replacements for these widely used compounds. Four distinct aspects related to the effectiveness of potential thermal agents have been considered. First, existing thermodynamic databases maintained by NIST have been searched in order to identify chemical compounds which are predicted to extract large amounts of heat from a combustion zone. Second, two liquids that were identified as being particularly interesting during the database search were tested experimentally for their extinguishing effectiveness. Third, detailed chemical-kinetic modeling and experimental studies of extinguishing volume fractions in simple flame systems were employed to improve the understanding of the effects of thermal agents on diffusion flames. Fourth, empirical heat transfer correlations for spray cooling of a surface were used to estimate the efficiencies of surface cooling by thermal agents. The database searches used two primary sources -- the Design Institute for Physical Properties database containing 1458 compounds from 83 family types and a smaller database, REFPROP, containing 43 compounds which is tailored to refrigerant applications. Additional substances were included that are not well represented in these databases. Compounds having high 1) heats of vaporization, 2) liquid-phase heat capacities, and 3) total heat absorption due to phase changes (if applicable), heating of a liquid (if applicable), and the heating of the gas phase to combustion temperatures were identified. The results are reported in tables of compounds ordered in terms of their ability to extract heat. The following recommendations are made based on the findings of the study: 1. The effectiveness of HFE7100 as a fire-extinguishing agent should be tested on full-scale simulations of aircraft fires. 2. Additional studies of the effectiveness of fire extinguishing agents released as liquids should be performed with a focus on confirming and understanding their enhanced performance relative to that expected based on simple heat extraction. 3. Experiments should be designed and performed with a goal of better understanding the interactions of liquid agent droplets with heated and reacting surfaces. Such an understanding is needed to better predict the effectiveness of liquid agents and could lead to improved approaches for delivering such agents to a fire.

Citation: NIST Interagency/Internal Report (NISTIR) - 1440

Keyword: fire extinguishing agents; reaction kinetics; flame extinguishment; surface cooling; combustion; propane; experiments; thermophysical properties; lactic acid; halon alternatives; diffusion flames; methane; extinction; argon; nitrogen; carbon dioxide; helium; water vapors; dilution; heat extraction; flame fronts; extinguishing; halon 1301

NIST Technical Note 1440

Characteristics and Identification of Super-Effective Thermal Fire-Extinguishing Agents: Final Report, NGP Project 4C/1/890

William M. Pitts
Jiann C. Yang
Rodney A. Bryant
Linda G. Blevins
Marcia L. Huber

National Institute of Standards and Technology
Technology Administration, U.S. Department of Commerce

NIST Technical Note 1440

Characteristics and Identification of Super-Effective Thermal Fire-Extinguishing Agents: Final Report, NGP Project 4C/1/890

William M. Pitts
Jiann C. Yang
Rodney A. Bryant
Linda G. Blevins
Building and Fire Research Laboratory

Marcia L. Huber
Chemical Science and Technology Laboratory

June 2001
Issued July 2006

U.S. Department of Commerce
Donald L. Evans, Secretary

National Institute of Standards and Technology
Dr. Karen H. Brown, Acting Director

Certain commercial entities, equipment, or materials may be identified in this document in order to describe an experimental procedure or concept adequately. Such identification is not intended to imply recommendation or endorsement by the National Institute of Standards and Technology, nor is it intended to imply that the entities, materials, or equipment are necessarily the best available for the purpose.

National Institute of Standards and Technology Technical Note 1440
Natl. Inst. Stand. Technol. Tech. Note 1440, 138 pages (July 2006)
CODEN: NSPUE2

TABLE OF CONTENTS

LIST OF TABLES..iii

LISTS OF FIGURES...v

EXECUTIVE SUMMARY..xi

I. INTRODUCTION..1
II. IDENTIFICATION AND TESTING OF POTENTIAL HIGHLY EFFECTIVE THERMAL AGENTS..2
   A. Screening Based on Thermophysical Properties..2
   B. Experimental Testing of Lactic Acid and HFE7100..11
      1. Lactic Acid..11
         a) Estimation of the latent heat of vaporization for lactic acid/water mixtures....11
         b) Experimental system..13
         c) Experimental results for lactic acid/water mixture fire suppression effectiveness............14
      2. HFE7100...16
         a) Testing of HFE7100 using the Dispersed Liquid Agent Fire Suppression Screen facility..16
         b) Testing of HFE7100 using the Transient Application Recirculating Pool Fire facility.....16
III. DETAILED CHEMICAL-KINETIC MODELING AND EXPERIMENTAL INVESTIGATIONS OF THEMAL AGENT FLAME EXTINGUISHMENT..19
   A. Introduction...19
   B. Introduction to Detailed Chemical-Kinetic Modeling of Diffusion Flames...............22
   C. Modeling Approach..27
      1. CHEMKIN III and OPPDIF..27
      2. Detailed Chemical-Kinetic Mechanism for Methane...27
      3. Detailed Chemical-Kinetic Mechanism for Propane..27
      4. Modeling Details..28
   D. Model Results for Methane and Propane Flames Burning in Air................................29
   E. Extinction and Extinguishment of Methane and Propane Diffusion Flames by Nitrogen.............38
      1. Experimental Systems for Measuring Extinguishing Concentrations of Gaseous Thermal Agents..40
         a) Modified Dispersed Liquid Agent Fire Suppression Screen System (Tsuji Burner)........40
         b) NIST Coflow Diffusion Flame Burner (Santoro Burner)...............................41
      2. Results for Thermal Agent Extinction of Methane and Propane Diffusion Flames Burning on Tsuji and Santoro Burners..43
         a) Tsuji Burner..43
         b) Santoro Burner..44
      3. Modeling of Extinction and Extinguishment of Methane and Propane Flames by Nitrogen...51
   F. Effects of Argon Addition on Extinction and Extinguishment of Methane Diffusion Flames Burning in Argon "Air"..56
   G. Extinguishment of Methane and Propane Diffusion Flames by Carbon Dioxide, Argon, Helium and Water Vapor...57
   H. Effects of Dilution and Heat Extraction Location Relative to the Flame Front on Extinction and Extinguishment of Diffusion Flames...70
   I. Prediction of Extinguishing Concentration for HFE7100..80
IV. SURFACE COOLING..81
   A. Introduction...81

	B. Evaporation of Liquid Droplets..83	
	C. Spray Cooling..86	
	D. Convective Regime...87	
	E. Nucleate Boiling Regime..87	
	F. Critical Heat Flux..88	
	G. Transition and Film Boiling Regimes..89	
	H. Summary...90	
V.	DISCUSSION..91
	A. Liquid HFE7100 As a Potential Fire-Fighting Agent and Liquid Agent Effectiveness................91
	B. Conclusions from Detailed Chemical-Kinetic Modeling of Thermal Agent Effects....................93
VI.	RECOMMENDATIONS FOR ADDITIONAL STUDY..96
VII.	ACKNOWLEDGEMENTS...96
VIII.	APPENDIX A--DETAILED CHEMICAL-KINETIC MECHANISM FOR PROPANE COMBUSITON....................97
IX.	APPENDIX B—ADDITIONAL DETAILS CONCERNING THE COFLOW DIFFUSION FLAME EXPERIMENT.................109
	A. Mass-Flow Controllers and Their Calibration..109
	B. Detailed Description of Gas Flow Calculations...109
	C. Uncertainty Analysis..113
X.	REFERENCES..115

LIST OF TABLES

Table 1. Family Types in the DIPPR Database..2

Table 2. Results of Search for Compounds with High Heat of Vaporization at the Normal Boiling Point........4

Table 3. References for Property Data (Critical Point, Normal Boiling Point, Heat of Vaporization, Liquid Heat Capacity)..5

Table 4. Results of Search for Compounds with High Liquid Phase Heat Capacities at the Normal Boiling Point..6

Table 5. Results of Search for Compounds with High ΔH^{Total} (Molar Basis)..................................7

Table 6. Results of Search for Compounds with High ΔH^{Total} (Mass Basis)...................................8

Table 7. Some Properties of Lactic Acid Taken From Reference [9]..11

Table 8. Calculated Equivalent Mass and Mole Fraction Values for the Reference Blow-Off Velocity..........16

Table 9. Some Parameters Used in Most OPPDIF Calculations of Methane and Propane Flames.................30

Table 10. Various Measures of Strain Rate Calculated for Methane/Air Diffusion Flames at Extinction.......34

Table 11. Extinguishment Results Using the DLAFSS..44

Table 12. Measured Agent Volume Fractions at Onset of Flame-Base Instability (Osc) and for Flame Extinguishment (Ext) of Baseline Methane and Propane Coflow Diffusion Flames.....................48

Table 13. Effects of Fuel and Oxidizer Velocity on Agent Concentrations at Onset of Flame Base Oscillation (Osc) and Flame Extinguishment (Ext) for the Methane Coflow Flame......................48

Table 14. Effects of Fuel and Oxidizer Velocity on Agent Concentrations at Onset of Flame Base Oscillation (Osc) and Flame Extinguishment (Ext) for the Propane Coflow Flame......................48

Table 15. Extinguishing Nitrogen Volume Fractions For Various Diffusion Flames Burning Methane and Propane..51

Table 16. Extinguishing Volume Fractions of Thermal Agents Calculated For Opposed-Flow Methane Flames and Measured For Heptane Flames in Cup Burner Experiments.........................60

Table 17. Calculated Extinguishing Volume Fractions for Methane and Propane Opposed-Flow Diffusion Flames for the Indicated Thermal Agents..67

Table 18. Adiabatic Flame Temperatures Calculated for Methane and Propane Premixed Flames Burning in Air Diluted with Extinguishing Volume Fractions of Thermal Agents.....................68

Table 19. Mass-Flow Controller Specifications..110

Table 20. Calibration Coefficients for the Air and Fuel Mass-Flow Controllers..................................110

Table 21. Calibration Coefficients for Path 1 Agent Mass-Flow Controller...110

Table 22. Calibration Coefficients for Path 2 Agent Mass-Flow Controller...110

Table 23. Agent Extinguishing Concentration Measurement Standard Deviation...114

Table 24. Uncertainty Budget – Agent Extinguishing Volume Fraction...114

LIST OF FIGURES

Figure 1. The relative contributions of liquid heating, vaporization, and gas heating to 1400 K to the total heat absorbed by various thermal agents on a per mole basis are shown as bar graphs..................9

Figure 2. The relative contributions of liquid heating, vaporization, and gas heating to 1400 K to the total heat absorbed by various thermal agents on a per mass basis are shown as bar graphs..................10

Figure 3. An equilibrium T-x,y diagram for lactic acid/water mixtures is shown for a pressure of 101 kPa. The symbols refer to the liquid ( ) and gaseous ( ) phases..........................12

Figure 4. Enthalpies for the saturated vapor at the dew point ( ) and saturated liquid at the bubble point ( ) are plotted as a function of water mole fraction for lactic acid/water mixtures..................12

Figure 5. The enthalpy of vaporization for lactic acid/water mixtures for a pressure of 101 kPa is plotted as a function of the water mole fraction..................13

Figure 6. A schematic of the dispersed liquid agent fire suppression screen (DLAFSS) apparatus is shown..................14

Figure 7. Velocities of the air at extinguishment (indicated by blow off of the opposed-jet diffusion flame) are plotted as a function of liquid application rate for the indicated lactic acid/water mixtures..................15

Figure 8. Velocities of the air at extinguishment (indicated by blow off of the opposed-jet diffusion flame) are plotted as a function of liquid application rate for HFE7100..................17

Figure 9. A schematic is shown for the TARPF facility [37,38] as modified for the investigation of the effectiveness of liquid fire suppression agents..................18

Figure 10. Maximum flame temperatures are plotted as a function of the equal fuel and air velocity magnitudes for a methane/air opposed-flow diffusion flame..................30

Figure 11. Calculated strain rates, a, are plotted as a function of distance from the fuel exit for low exit velocity magnitudes (25 cm/s for fuel and air) and for a condition very close to extinction (320.1 cm/s)..................31

Figure 12. The maximum flame temperature is plotted against a characteristic local strain rate, $|a_o|$, defined to be equal to the highest absolute value of a on the air side observed outside of the boundary layer associated with the methane/air flame..................32

Figure 13. Mixture fraction is plotted as a function of distance from the fuel exit for low exit velocity magnitudes (25 cm/s for fuel and air) and for a condition very close to extinction (320.1 cm/s) for a methane/air diffusion flame..................32

Figure 14. Calculated scalar dissipation rates, χ, are plotted as a function of distance from the fuel exit for low exit velocity magnitudes (25 cm/s for fuel and air) and for a condition very close to extinction (320.1 cm/s) for a methane/air diffusion flame..................33

Figure 15. Values of maximum flame temperature for opposed-flow methane/air diffusion flames are plotted as a function of χ_s that is varied by changing the velocities of the fuel and air. Extinction is calculated to occur for $\chi_s = 20.7$ s^{-1}..................33

Figure 16. Maximum flame temperatures are plotted as a function of the equal fuel and air velocity magnitudes for a methane/air opposed-flow diffusion flame. Results are compared for calculations using Versions 1.2, 2.11, and 3.0 of the GRIMech mechanism. [106]..................35

Figure 17. Maximum flame temperatures are plotted as a function of the equal fuel and air velocity magnitudes for a propane/air opposed-flow diffusion flame..................36

Figure 18. The maximum flame temperature is plotted as a function of the global stain rate a_g for a propane/air opposed-flow diffusion flame..................36

Figure 19. The maximum flame temperature is plotted against a characteristic local strain rate, $|a_o|$, defined to be equal to the highest absolute value of a on the air side observed outside of the boundary layer associated with the combustion zone for the propane/air flame..................37

Figure 20. Values of maximum flame temperature for opposed-flow propane/air diffusion flames are plotted as a function of χ_s that is varied by changing the velocities of the fuel and air. Extinction is calculated to occur for $\chi_s = 26.3$ s^{-1}..................38

Figure 21. A schematic of the modified Santoro burner is shown..................41

Figure 22. A schematic and photo of the Santoro burner and flow control system are shown..................42

Figure 23. Sketch of a methane flame showing the attached flame oscillation and subsequent stabilized lifted flame that result from the addition of nitrogen to the air coflow..................45

Figure 24. The evolution of the methane diffusion flame structure as additional N_2 is added to the coflow air stream is shown..................45

Figure 25. A sketch of the time evolution of a nearly extinguished propane flame is shown for the case of nitrogen dilution of the coflow air, $t_1 < t_3$..................46

Figure 26. A time-resolved photograph taken by a digital camera shows the oscillating propane flame burning in air diluted with nitrogen. The complex azimuthal structure of the flame base, which has a much larger area than the burner tube, is clear..................49

Figure 27. The cartoon shows the time evolution of the pulsing instability in the large flow-velocity propane diffusion flame observed when a concentration of a thermal fire-extinguishing agent near that required for extinguishment is added to the air coflow. The arrows indicate the direction of flame base propagation. The view is from an angle above the flame..................50

Figure 28. Maximum calculated flame temperatures are plotted against methane and oxidizer exit velocity magnitudes for a series of methane/air opposed-flow diffusion flames diluted with nitrogen. Symbols correspond to the volume fraction of added nitrogen..................52

Figure 29. Maximum calculated flame temperatures are plotted against propane and oxidizer exit velocity magnitudes for a series of propane/air opposed-flow diffusion flames diluted with nitrogen. Symbols correspond to the volume fraction of added nitrogen..................52

Figure 30. Values of the maximum calculated flame temperature at extinction are plotted as a function of the fuel and oxidizer flow velocities for methane and propane flames burning in air diluted by various volume fractions of nitrogen..53

Figure 31. Values of maximum flame temperatures calculated using OPPDIF are plotted as a function of added nitrogen volume fraction for methane and propane opposed-flow flames. These values are compared with the corresponding adiabatic flame temperatures assuming stoichiometric combustion..54

Figure 32. Ratios for the indicated measures of conditions at flame extinction are plotted as a function of nitrogen volume fraction added to the air for methane/air flames...55

Figure 33. Maximum calculated flame temperatures are plotted against methane and oxidizer flow velocity magnitudes for a series of methane/argon "air" opposed-flow diffusion flames diluted with argon. Symbols correspond to the volume fraction of added argon.....................................57

Figure 34. Maximum calculated flame temperatures are plotted against methane and oxidizer flow velocity magnitudes for a series of methane/air opposed-flow diffusion flames diluted with argon. Symbols correspond to the volume fraction of added argon...58

Figure 35. Maximum calculated flame temperatures are plotted against methane and oxidizer flow velocity magnitudes for a series of methane/air opposed-flow diffusion flames diluted with helium. Symbols correspond to the volume fraction of added helium...58

Figure 36. Maximum calculated flame temperatures are plotted against methane and oxidizer flow velocity magnitudes for a series of methane/air opposed-flow diffusion flames diluted with carbon dioxide. Symbols correspond to the volume fraction of added carbon dioxide................59

Figure 37. Maximum calculated flame temperatures are plotted against methane and oxidizer flow velocity magnitudes for a series of methane/air opposed-flow diffusion flames diluted with water vapor. Symbols correspond to the volume fraction of added water vapor..........................59

Figure 38. Detailed chemical-kinetic modeling predictions for extinguishing volume fractions of nitrogen, carbon dioxide, argon, and helium for methane flames are compared with various experimental cup burner measurements using n-heptane. Model values for helium and nitrogen overlap and open symbols are used to identify the helium results. References for the cup burner measurements are Hamins et al. [33], Hirst and Booth [47], Sheinson et al. [49], Moore et al. [54], Saito et al. [55], and Babb et al. [56]..61

Figure 39. Values of maximum flame temperatures (solid symbols) calculated using OPPDIF and the corresponding adiabatic flame temperatures (open symbols) calculated using EQUIL for methane flames are plotted as a function of added agent volume fraction in air for five thermal agents. Note that the adiabatic flame temperatures for argon and helium overlap and that the maximum calculated flame temperatures for helium and nitrogen fall very close together..........62

Figure 40. Values of maximum flame temperature at extinction are plotted as a function of the corresponding fuel and oxidizer velocity magnitudes for a range of volume fractions of the various thermal agents..62

Figure 41. Values of maximum flame temperature at extinction are plotted as a function of the corresponding global strain rates, a_g, for a range of volume fractions of the various thermal agents...63

Figure 42. Values of maximum flame temperature at extinction are plotted against the corresponding characteristic local strain rates, $|a_o|$, defined as the highest absolute value of strain rate observed outside of the boundary layer on the oxidizer side of the flame, for a range of volume fractions of the various thermal agents...64

Figure 43. Maximum flame temperatures at extinction for N_2, Ar, He, and CO_2 as a function of the corresponding fuel and oxidizer velocity magnitudes are fit by two piecewise fourth-order polynomials (solid line)...65

Figure 44. The maximum temperature at extinction for opposed-flow methane flames is plotted as a function of the volume fraction of a thermal agent added to air divided by the volume fraction calculated to be required for flame extinguishment...65

Figure 45. Calculated maximum flame temperatures are plotted as a function of velocity for a propane flame burning in air diluted with the indicated volume fractions of carbon dioxide. The extinguishing volume fraction is defined to be that (28.4 %) which yields an extinction temperature of 1433 K..66

Figure 46. Calculated maximum flame temperatures are plotted as a function of velocity for a propane flame burning in air diluted with the indicated volume fractions of argon. The extinguishing volume fraction is defined to be that (50.2 %) which yields an extinction temperature of 1433 K..66

Figure 47. Calculated maximum flame temperatures are plotted as a function of velocity for a propane flame burning in air diluted with the indicated volume fractions of helium. The extinguishing volume fraction is defined to be that (39 %) which yields an extinction temperature of 1433 K. 67

Figure 48. Experimental values of extinguishing volume fractions for N_2, CO_2, Ar, and He for counterflow porous (circles), opposed-jet (squares) and coflow (triangles) burners are plotted against the corresponding OPPDIF predictions (dotted line) for methane-air diffusion flames. Experimental values are taken from the present work as well from the following references: Ishizuka and Tsuji [48], Simmons and Wolfard [41], Puri and Seshadri [114], Hamins [121], and Ural [57]. Current results for helium are indicated by open symbols.....................68

Figure 49. Experimental values of extinguishing concentrations for N_2, CO_2, Ar, and He for counterflow porous (circles), opposed-jet (squares) and coflow (triangles) burners are plotted against the corresponding OPPDIF predictions (dotted line) for propane-air diffusion flames. Experimental values are taken from the present work as well from the following references: Simmons and Wolfhard [41], Puri and Seshadri [114], Hamins [121], Ural [57], Babb et al. [56], Moore et al. [54], and Hamins et al. [33]..69

Figure 50. Maximum calculated flame temperatures are plotted against methane and oxidizer exit velocity magnitudes for a series of methane/air opposed-flow diffusion flames diluted with a surrogate agent similar to argon, but with $C_p = 0$. Symbols correspond to the volume fraction of surrogate agent added to the air...70

Figure 51. Values of maximum flame temperature at extinction are plotted as a function of fuel and oxidizer extinction velocity magnitudes for a range of concentrations of argon and a similar surrogate agent having $C_p = 0$. The solid line is the result of the regression fits shown in Figure 43..................71

Figure 52. Temperature is plotted as a function of distance from the fuel exit for methane flames burning in air and air mixed with 60 % of a diluent having similar properties to Ar, but with zero heat capacity. Fuel and air exit velocity magnitudes are 25 cm/s for both flames..................72

Figure 53. Calculated methane and oxygen volume fractions are shown on a semi-log plot as a function of distance from the fuel exit for methane flames burning in air (open symbols) and air mixed with 60 % of a diluent having similar properties to Ar, but with zero heat capacity. Fuel and air exit velocity magnitudes are 25 cm/s for both flames..................72

Figure 54. Calculated flame temperature is plotted as a function of distance from the fuel exit for methane/air diffusion flames where the oxidizer is diluted with 5 % of either argon or the hypothetical agent X which reacts to form Y with $E_a = 25.1$ kJ/mol and $\Delta H_{X \to Y} = 96.1$ kJ/mol. Methane and oxidizer exit velocity magnitudes are 25 cm/s..................74

Figure 55. The volume fractions of X and Y are plotted as a function of distance from the fuel exit for a methane flame burning in air with 5 % added X. X reacts to form Y with $E_a = 25.1$ kJ/mol and $\Delta H_{X \to Y} = 96.1$ kJ/mol. The spatial variation of temperature is also shown as the solid line. Methane and oxidizer exit velocity magnitudes are 25 cm/s..................74

Figure 56. The volume fractions of X and Y are plotted as a function of distance from the fuel exit for a methane flame burning in air with 5 % added X. X reacts to form Y with $E_a = 58.6$ kJ/mol and $\Delta H_{X \to Y} = 96.1$ kJ/mol. The spatial variation of temperature is also shown as the solid line. Methane and oxidizer exit velocity magnitudes are 25 cm/s..................75

Figure 57. The volume fractions of X and Y are plotted as a function of distance from the fuel exit for a methane flame burning in air with 5 % added X. X reacts to form Y with $E_a = 83.7$ kJ/mol and $\Delta H_{X \to Y} = 96.1$ kJ/mol. The spatial variation of temperature is also shown as the solid line. Methane and oxidizer exit velocity magnitudes are 25 cm/s..................76

Figure 58. Values of maximum flame temperature for methane flames burning in air diluted with 5 % X are plotted as a function of the energy of activation for the conversion of X to Y. Methane and oxidizer exit velocity magnitudes are 25 cm/s..................77

Figure 59. Maximum calculated flame temperatures are plotted against methane and oxidizer exit velocity magnitudes for a series of methane/air opposed-flow diffusion flames diluted with agent X that reacts to form Y with $E_a = 41.8$ kJ/mol and $\Delta H_{X \to Y} = 96.1$ kJ/mol. Symbols correspond to the volume fractions of added X..................77

Figure 60. Maximum calculated flame temperatures are plotted against methane and oxidizer exit flow velocity magnitudes for a series of methane/air opposed-flow diffusion flames diluted with argon and agent X ($E_a = 41.8$ kJ/mol and $\Delta H_{X \to Y} = 96.1$ kJ/mol)..................78

Figure 61. Maximum calculated flame temperatures are plotted against methane and oxidizer exit flow velocity magnitudes for a series of methane/air opposed-flow diffusion flames diluted with agent X which reacts to form Y with $E_a = 41.8$ kJ/mol and $\Delta H_{X \to Y} = 192.1$ kJ/mol. Symbols correspond to the volume fractions of added X..................79

Figure 62. Values of maximum flame temperature at extinction are plotted as a function of fuel and oxidizer exit velocity magnitudes for air diluted with a range of volume fractions of argon and surrogate agent X having $\Delta H_{X \to Y}$ = 96.1 kJ/mol or 192.1 kJ/mol. The solid line is the result of the regression fits shown in Figure 43 79

Figure 63. Maximum calculated flame temperatures are plotted against methane and oxidizer exit velocity magnitudes for a series of methane/air opposed-flow diffusion flames diluted with $C_4F_9OCH_3$. Symbols correspond to the volume fractions of added $C_4F_9OCH_3$ 80

Figure 64. The droplet evaporation time is shown as a function of initial droplet diameter for the five fluids indicated 85

Figure 65. An illustration of a typical boiling curve associated with quenching of a hot surface by liquid droplets is shown 86

Figure 66. Calculated boiling curves in the convective regime are shown for the five indicated liquids 88

Figure 67. Calculated boiling curves for the five liquids are shown for the transition and film boiling regimes 90

EXECUTIVE SUMMARY

The use of halon 1301 for fire fighting is being phased out due to its deleterious effects on stratospheric ozone. This report summarizes the findings of a three-year study designed to characterize and identify super-effective thermal fire-fighting agents as possible replacements for these widely used compounds. Four distinct aspects related to the effectiveness of potential thermal agents have been considered. First, existing thermodynamic databases maintained by NIST have been searched in order to identify chemical compounds which are predicted to extract large amounts of heat from a combustion zone. Second, two liquids that were identified as being particularly interesting during the database search were tested experimentally for their extinguishing effectiveness. Third, detailed chemical-kinetic modeling and experimental studies of extinguishing volume fractions in simple flame systems were employed to improve the understanding of the effects of thermal agents on diffusion flames. Fourth, empirical heat transfer correlations for spray cooling of a surface were used to estimate the efficiencies of surface cooling by thermal agents.

The database searches used two primary sources—the Design Institute for Physical Properties database containing 1458 compounds from 83 family types and a smaller database, REFPROP, containing 43 compounds which is tailored to refrigerant applications. Additional substances were included that are not well represented in these databases. Compounds having high 1) heats of vaporization, 2) liquid-phase heat capacities, and 3) total heat absorption due to phase changes (if applicable), heating of a liquid (if applicable), and the heating of the gas phase to combustion temperatures were identified. The results are reported in tables of compounds ordered in terms of their ability to extract heat.

Two compounds, methoxy-nonaflurobutane (HFE7100) and lactic acid, were identified during the data base searches as being especially interesting. HFE7100 was predicted to be particularly effective as a thermal agent and is available commercially. Lactic acid is capable of absorbing large quantities of heat even though it is very caustic and is flammable at certain volume fractions when mixed with air. Both agents were tested in the Dispersed Liquid Agent Fire Suppression Screen (DLAFSS) apparatus available at NIST. The effective extinguishing volume fraction of HFE7100 when released as a liquid was 1.5 %. Lactic acid was tested in mixtures with water at various concentrations. The acid mixtures were found to be less effective than water alone, indicating that the lactic acid was reacting and releasing heat that more than compensated for the heat extraction.

HFE7100 was also tested in a screening apparatus known as the Transient Application Recirculating Pool Fire (TARPF) that incorporates a flame stabilized behind a bluff body in a turbulent oxidizer flow. The agent failed to extinguish a propane flame when it was released at nominal volume fractions in air twice as high as the extinguishing volume fraction measured in the DLAFSS. This observation is attributed to ineffective mixing.

Even though the effects of thermal agents on diffusion flames have been widely studied, there are relatively few experimental or modeling investigations of the flame extinguishment process. As a result, flame extinguishment by thermal agents is incompletely understood. A combined experimental and modeling study was used to address this concern. Careful experimental measurements of extinguishing volume fractions by gaseous thermal agents added to air were made for two types of diffusion flames available at NIST. The first was a coflow laminar flame that has a similar configuration as the cup burner, which is widely used to test fire-fighting agents, and the second was the modified DLAFSS apparatus that incorporates a cylindrical porous burner with the fuel released in the opposite direction of the oxidizer flow (i.e., an opposed flow burner). Measurements were made for two fuels: methane and propane, and four thermal agents: nitrogen, carbon dioxide, argon, and helium. The results showed that the counterlfow flame was more difficult to extinguish than the coflow flame and that propane flames required higher volume fractions of an agent for extinguishment than methane flames. The relative effectiveness of the agents in volume fraction terms was argon < nitrogen ≅ helium < carbon dioxide.

Detailed chemical-kinetic modeling of opposed-flow diffusion flames was used to obtain insights into the effectiveness of thermal agents and their mechanisms of flame extinction. This approach is based on the hypothesis that the maximum flame temperature at extinguishment for a given fuel is a unique value as long as the agent is nonreactive in the flame. This hypothesis was tested by using experimental values for the volume fractions of nitrogen required to extinguish porous burner counterflow flames as the basis for the determination of the maximum flame temperature at extinguishment. This temperature was then used to determine the predicted extinguishing volume fractions for other agents. The calculations were performed for nitrogen, carbon dioxide, argon, helium, and water vapor added to the air and for methane and propane flames. For the methane flame the maximum flame temperature at extinguishment was calculated to be 1550 K and for the propane flame it was 1433 K.

Comparisons of the predicted extinguishing volume fractions with the experimental measurements indicated that the chemical-kinetic modeling was capable of accurately predicting extinguishing conditions. The results for the methane flame were also found to be in good agreement with numerous published measurements of thermal agent extinguishing volume fractions for heptane pool fires recorded in cup burner experiments by a number of researchers. In particular, the calculations reproduced the experimental observation that similar volume fractions of nitrogen and helium are required to extinguish a flame, despite the fact that the heat capacity for nitrogen is considerably greater. The calculations also reproduced the experimental finding that propane flames are more difficult to extinguish than methane flames.

A series of calculations were performed for gaseous HFE7100. An extinguishing concentration of 5.5 % was predicted, which is in good agreement with two unpublished experimental values of 6.1 %. Note that these values are nearly four times greater than the experimental value recorded when HFE7100 was released as a liquid. Measurements by other workers have indicated that water released as a liquid is also more effective than is predicted by simply considering the effects of heat removal. At present, these observations are not fully understood, but they do suggest that liquid agents may gain effectiveness from additional mechanisms.

An advantage of detailed chemical-kinetic modeling studies is that surrogate agents having properties that are not physically realizable can be used to investigate specific details concerning extinguishment. A surrogate agent was specified which reacted over different temperature ranges to extract a predetermined amount of heat. The calculations showed that the effectiveness of this agent was independent of the location of heat extraction relative to the flame zone. In a second series of calculations a surrogate agent was used to isolate the role of dilution on extinguishment. When the agent, which was incapable of extracting heat, was added to the air, much higher concentrations were required to extinguish the flame than when heat was extracted. Details of the calculations revealed that extinguishment ultimately occurred due to oxygen passing through the flame zone as a result of the decreased overall reaction rate.

Calculations of droplet evaporation times using the classical d^2-law for the five fluids (water, lactic acid, $C_3F_5H_3O$, HFE7100, and R338mccq) identified as having the highest latent heat of vaporization (per unit mass) by the database searches were performed as part of the surface cooling studies. Empirical heat transfer correlations from the spray surface quenching literature were used to assess the surface cooling characteristics of these fluids for various heat transfer regimes. Based on these calculations, water and lactic acid appear to be more effective than the other three fluids for surface cooling applications.

The following recommendations are made based on the findings of the study:

1. The effectiveness of HFE7100 as a fire-extinguishing agent should be tested on full-scale simulations of aircraft fires.

2. Additional studies of the effectiveness of fire extinguishing agents released as liquids should be performed with a focus on confirming and understanding their enhanced performance relative to that expected based on simple heat extraction.

3. Experiments should be designed and performed with a goal of better understanding the interactions of liquid agent droplets with heated and reacting surfaces. Such an understanding is needed to better predict the effectiveness of liquid agents and could lead to improved approaches for delivering such agents to a fire.

I. INTRODUCTION

The use of halon 1301 for fire fighting is being phased out due to its deleterious effects on stratospheric ozone. This report summarizes the findings of a three-year study designed to characterize and identify super-effective thermal fire-fighting agents as possible replacements for these effective compounds. "Thermal" agents refer to compounds that act simply by extracting heat from a flame zone and lowering the temperature to a point where combustion can no longer be sustained. These types of agents should be contrasted with "chemical" agents that generate active chemical species (e.g., the bromine atoms generated by halons) that interfere with the radical chain branching mechanisms required to sustain combustion.

There are a number of endothermic physical processes that can extract heat from a gaseous flame zone, thus lowering the temperature and ultimately leading to flame extinguishment. These include simple heating (i.e., heat capacity) of an agent, phase changes such as vaporization of a liquid or sublimation of a solid, endothermic molecular decomposition (note that this process is classed as a physical process as long as the initial agent and its products do not participate in the combustion chemistry), and simple dilution which can modify flame temperatures by spreading the heat released by combustion over larger volumes and by slowing three-body reactions due to lower species collision rates. The flame temperature is also expected to be a function of the thermal diffusivity of an agent.

Another potentially effective way for thermal agents to extinguish a flame fueled by liquids or solids is to interact with the fuel surface to lower the temperature to a point where the fuel generation rate is insufficient to maintain combustion. This mechanism for fire extinguishment has not been extensively studied, and the physical properties that determine the effectiveness of an agent operating in this mode are not known.

During this investigation several aspects related to the effectiveness of potential thermal agents have been considered. One part of the study focused on identifying and testing chemicals likely to be particularly effective as thermal fire extinguishing agents. Existing thermodynamic data bases maintained by NIST were searched to select compounds predicted to absorb large amounts of heat as the temperature is raised. Two chemicals—lactic acid and methoxynonafluorobutane—were identified as being particularly interesting. The effectiveness of these chemicals were characterized using screening tests developed as part of the Next Generation Fire Suppression Technology Program (NGP) of the Strategic Environmental Research and Development Program (SERDP) of the U. S. Department of Defense. A combined modeling and experimental effort was aimed at improving the understanding of the effects of thermal agents on fires. This effort was motivated in part by the realization that different types of experiments yield significantly different estimates for extinguishing concentrations. The effectiveness of the well-known thermal agents nitrogen, carbon dioxide, argon, helium, and water were emphasized. Fuel effects were also considered. Lastly, empirical heat transfer correlations for spray cooling of a surface have been used to estimate the efficiencies of surface cooling by thermal agents that are expected to be very efficient at extracting heat from combustion regions.

Separate major sections summarize the findings of the three major parts of the study. A fourth section summarizes the major findings and provides additional discussion, and the final section provides recommendations for additional research and development. This report incorporates material that is included in an earlier National Institute of Standards and Technology Internal Report that summarized the findings of the first year of the effort. [1] Selected aspects have been presented during a number of technical meetings. [2,3,4,5,6,7,8]

Table 1. Family Types in the DIPPR Database

1 n-Alkanes	29 Cycloaliphatic alcohols	57 Other amines, imines
2 Methylalkanes	30 Aromatic alcohols	58 Nitriles
3 Dimethylalkanes	31 Polyols	59 C,H,NO2 compounds
4 Other alkanes	32 n-Aliphatic acids	60 Isocyanates/diisocyanates
5 Cycloalkanes	33 Other aliphatic acids	61 Mercaptans
6 Alkylcyclopentanes	34 Dicarboxylic acids	62 Sulfides/theophenes
7 Alkylcyclohexanes	35 Aromatic carboxylic acids	63 Polyfunctional acids
8 Multi-ring cycloalkanes	36 Anhydrides	64 Polyfunctional esters
9 1-Alkynes	37 Formates	65 Other polyfunctional C,H,O
10 2,3,4-Alkenes	38 Acetates	66 Polyfunctional nitriles
11 Methylalkenes	39 Propionates & Butyrates	67 Nitroamines
12 Ethyl and higher alkenes	40 Other saturated aliphatic esters	68 Polyfunc. amides/amines
13 Cycloalkenes	41 Unsaturated aliphatic esters	69 Polyfunctional C,H,O,N
14 Dialkenes	42 Aromatic esters	70 Polyfunctional C,H,O,S
15 Alkynes	43 Aliphatic ethers	71 Polyfunc. C,H,O,halides
16 n-Alkylbenzenes	44 Other ethers and diethers	72 Polyfunc. C,H,N,halides
17 Other alkylbenzenes	45 Epoxides	73 Other polyfunc. organics
18 Other monoaromatics	46 Peroxides	74 Elements
19 Napthalenes	47 C1/C2 Aliphatic chlorides	75 Silanes/siloxanes
20 Other condensed rings	48 C3 & higher aliph. chlorides	76 Organic-inorganic
21 Diphenyl/polyaromatics	49 Aromatic chlorides	77 Inorganic acids
22 Terpenes	50 C,H,Br compounds	78 Inorganic bases
23 Other hydrocarbon rings	51 C,H,I compounds	79 Organic salts
24 Inorganic gases	52 C,H,F compounds	80 Sodium salts
25 Aldehydes	53 C,H multihalogen compounds	81 Other inorganic salts
26 Ketones	54 n-Aliphatic primary amines	82 Inorganic halides
27 n-Alcohols	55 Other aliphatic amines	83 Other inorganics
28 Other alcohols (aliphatic)	56 Aromatic amines	

II. IDENTIFICATION AND TESTING OF POTENTIAL HIGHLY EFFECTIVE THERMAL AGENTS

The objective of this sub-task was to identify and test chemical compounds that were particularly promising as thermal agents. The first step in this process was to search thermophysical property databases and identify substances that had the potential to be particularly effective fire extinguishing agents based on thermophysical properties. Two compounds were subsequently tested experimentally using fire suppression screens developed as part of the SERDP effort.

A. Screening Based on Thermophysical Properties

We identified two thermophysical property databases to search. The first was the DIPPR (Design Institute for Physical Properties) Database, version 9.02 [9] which contained 1458 substances from 83 family types. The families are listed in Table 1. The second database, REFPROP, Version 5, [10] is a much smaller database tailored to alternative refrigerants. It contains 43 refrigerants including many candidates for replacing HCFCs. Some are fluorinated ethers (E134, E245) that are not present in the DIPPR database.

In addition, we wanted to include representative substances from some families of chemicals not well represented in existing thermophysical property databases, such as fluorinated ethers, alcohols and amines. The fluorinated ethers are especially interesting since their atmospheric lifetimes may be significantly lower than those of chlorofluorocarbons. [11] Some of these have been identified as possible replacements for blowing agents, refrigerants, fire suppression agents, and as solvents. [12,13,14,15,16,17] We selected several cyclic fluorinated ether compounds [fluorinated oxiranes, oxetanes, and furans], some non-cyclic methyl ethers [HFE116, HFE125], fluorinated methyl ethyl ethers with two or fewer hydrogen atoms [HFE227me, HFE236me], and some larger fluorinated ethers [perfluorodimethoxymethane ($CF_3OCF_2OCF_3$), perfluoropropyltrifluoromethyl ether ($CF_3OCF_2CF_2CF_3$), heptafluoropropyl-1,2,2,2-tetrafluoroethyl ether ($CF_3CHFOCF_2CF_2CF_3$), and methoxy-nonafluorobutane ($C_4F_9OCH_3$, HFE7100)]. In addition to the fluorinated ethers, we added two fluorinated amines to the list of substances to search and also the alternative refrigerant HFC-338mccq ($CH_2FCF_2CF_2CF_3$) and the fluorinated alcohol 2,2,3,3,3-pentafluoropropanol ($CF_3CF_2CH_2OH$).

The physical properties that are important for thermal agents are high latent heats of vaporization and high liquid and vapor heat capacities. We first searched the databases and the additional fluids mentioned above for high latent heats of vaporization. For fluids where data were unavailable, we estimated the critical properties using the NIST Structures and Properties program [18] and then estimated the latent heats using the approach of Fishtine [19]. We then eliminated substances on the list that either were identified as flammable, toxic, or had known ozone depletion problems. However, we did not have complete information on toxicity, so some substances that appear on our final list will need to be investigated further to determine if there are potential deleterious health effects. We also did not use global warming potential or electrical conductivity as screening parameters. If one considers global warming potential, then the fully fluorinated (perfluorinated) alkanes such as R218 and decafluorobutane would probably have to be ruled out. We used information in the DIPPR database on flash points and upper and lower flammability limits. The DIPPR database does not contain any information on toxicity or other health effects. For this information we used manufacturers safety data sheets and Hawley. [20] We also used the web site at http://chemfinder.camsoft.com/ that gives information on ozone depletion and global warming.

Table 2 lists the compounds that remained, in order of decreasing heat of vaporization at the normal boiling point, and includes R13B1 and water for reference. Table 3 identifies the sources of property data for critical points, normal boiling points, latent heat of vaporization, and liquid heat capacity. The general trend shown in Table 2 is quite predictable--the larger molecules have higher boiling points and higher heats of vaporization. Lactic acid is the only unusual fluid that appears is Table 2--all other fluids are basically {C, F, H}, {C, F} or {C, F, O} compounds. Lactic acid, $CH_3CH(OH)COOH$, is a strong irritant to the skin and corrosive in concentrated form and would not be a suitable agent. However, it may be at least partially soluble in fluorinated ethers such as HFE134 or in alternative refrigerants such as R227ea.

The second property we searched on was liquid heat capacity at the normal boiling point. When data were unavailable, the liquid heat capacity was estimated using the Chueh-Swanson method. [21] This method provides the liquid heat capacity at 20 EC based on the structure of a compound. The results are shown in Table 4 in order of decreasing liquid-phase heat capacity at the normal boiling point and include R13B1 and water for reference. Again, any compounds that were known to be flammable or toxic or to have known ozone depletion problems were eliminated. Table 4 indicates the same trend that was seen in Table 2, i.e., that larger compounds have higher boiling points and higher heats of vaporization and heat capacities. Also, in general, the ethers have higher boiling points and higher heats of vaporization at their normal boiling points than their alkane analogs. The fluorinated alcohol has a high boiling point due to the hydrogen bonding that occurs in alcohols.

Table 2. Results of Search for Compounds with High Heat of Vaporization at the Normal Boiling Point

Compound	T_b(K)	ΔH_{vap} (kJ/mol)
lactic acid, $C_3H_6O_3$	455	59.2
water	373.1	40.0
2,2,3,3,3-pentafluoropropanol, $C_3F_5H_3O$	353.15	37.6
perfluoro-2-butyltetrahydrofuran, $C_8F_{16}O$	375.15	35.4
methoxy-nonafluorobutane, HFE7100	334.0	30.3
heptafluoropropyl-1,2,2,2-tetrafluoroethyl ether	315.2	29.0
difluoromethyl-1,2,2,2-tetrafluoroethyl ether, E236me	296.5	26.9
1,1,1,2,2,3,3,4-octafluorobutane, R338mccq	301	26.8
tris(difluoromethyl)amine, $C_3H_3F_6N$	275	25.8
1,1,2,2-tetrafluorodimethylether, E134	279.4	25.4
1,1,1,2,3,3-hexafluoropropane, R236ea	279.7	25.1
perfluorotrimethylamine, C_3F_9N	264	23.9
trifluoromethyl-1,2,2,2-tetrafluoroethyl ether, E227me	263.6	23.7
octafluorocyclobutane, C_4F_8	267	23.2
decafluorobutane, C_4F_{10}	271	23.2
perfluorodimethoxymethane, $CF_3OCF_2OCF_3$	263	23.1
perfluoropropylmethyl ether, $C_3F_7OCF_3$	279.9	22.8
bis(trifluoromethyl)ether, E116	257	22.8
octafluorofuran, C_4F_8O	272.42	22.7
1,1,2,2-tetrafluoroethane, R134	250	22.5
1,1,1,2-tetrafluoroethane, R134a	247	22.3
perfluoroisobutane, C_4F_{10}	252.45	22.0
octafluoro-2-butene	270.4	21.7
1,1,1,2,3,3,3-heptafluoropropane, R227ea	257.5	21.4
perfluorooxetane, C_3F_6O	244.0	20.9
pentafluorodimethylether, E125	235	20.0
octafluoropropane, R218	236.4	19.7
hexafluoropropylene	243.5	19.4
pentafluoroethane, R125	225.2	19.1
bromotrifluoromethane, R13B1	215.26	17.4
hexafluoroethane, R116	194.95	16.1

In addition to looking at the latent heat of vaporization and the liquid heat capacity at the normal boiling point, one may also look at the total heat absorbed in going from the fluid at room temperature up to a temperature of 1400 K, which was selected as a representative combustion temperature. Other choices are possible including the calculated maximum flame temperatures of 1550 K and 1433 K identified below during the detailed chemical-kinetic modeling of methane and propane flames, respectively. The actual value used is unlikely to affect the relative ordering of agents. The total heat absorbed consists of three contributions,

$$\Delta H^{Total} = \Delta H_l + \Delta H_v + \Delta H_g, \tag{1}$$

Table 3. References for Property Data (Critical Point, Normal Boiling Point, Heat of Vaporization, Liquid Heat Capacity)

Compound	References
perfluoro-2-butyltetrahydrofuran, $C_8F_{16}O$ 22,	21, 19
methoxy-nonafluorobutane, HFE7100	23, 21, 19
lactic acid, $C_3H_6O_3$ 9	
heptafluoropropyl-1,2,2,2-tetrafluoroethyl ether	18, 21, 19
decafluorobutane, C_4F_{10} 9	
perfluoropropylmethyl ether, $C_3F_7OCF_3$ 24,	21, 19
perfluorodimethoxymethane, $CF_3OCF_2OCF_3$ 25,	21, 19
octafluoro-2-butene 9	
octafluorofuran, C_4F_8O	24, 21, 19
octafluorocyclobutane, C_4F_8 9	
perfluorotrimethylamine, C_3F_9N	18, 21, 19
1,1,1,2,3,3,3-heptafluoropropane, R227ea	10
perfluoroisobutane, C_4F_{10}	12, 21, 19
tris(difluoromethyl)amine, $C_3H_3F_6N$	18, 21, 19
trifluoromethyl-1,2,2,2-tetrafluoroethyl ether, E227me	26, 21, 19
1,1,1,2,2,3,3,4-octafluorobutane, R338mccq	27, 21, 19
1,1,1,2,3,3-hexafluoropropane, R236ea	10
difluoromethyl-1,2,2,2-tetrafluoroethyl ether, E236me	26, 21, 19
octafluoropropane, R218	10
2,2,3,3,3-pentafluoropropanol, $C_3F_5H_3O$ 28,	21, 19
perfluorooxetane, C_3F_6O	25, 21, 19
1,1,2,2-tetrafluorodimethylether, E134	29
bis(trifluoromethyl)ether, E116	18, 21, 19
hexafluoropropylene 9	
pentafluorodimethylether, E125	24, 21, 19
pentafluoroethane, R125	10
1,1,2,2-tetrafluoroethane, R134	10
hexafluoroethane, R116	10
1,1,1,2-tetrafluoroethane, R134a	10
bromotrifluoromethane, R13B1	1
water 9	

where the first contribtion, ΔH_1, is the heat involved in taking the liquid from room temperature up to the normal boiling point, T_b,

$$\Delta H_1 = \int_{298K}^{T_b} C_p^{liquid} \, dT. \quad (2)$$

Since the heat capacity of a liquid does not change significantly over a small temperature range, we used the value of liquid heat capacity at the normal boiling point. If this was unavailable, we estimated the liquid phase heat capacity at 20 EC using the method of Chueh and Swanson. [21] The second term, ΔH_v, is the latent heat of vaporization at the normal boiling point. The third contribution, ΔH_g, is found by integrating the gas-phase heat capacity from the normal boiling point to 1400 K,

Table 4. Results of Search for Compounds with High Liquid Phase Heat Capacities at the Normal Boiling Point

Compound	T_b (K)	C_p (J/mol K)
perfluoro-2-butyltetrahydrofuran, $C_8F_{16}O$ 375.15		364.2
methoxy-nonafluorobutane, HFE7100	334.0	292.8
lactic acid, $C_3H_6O_3$ 455		290.1
heptafluoropropyl-1,2,2,2-tetrafluoroethyl ether	315.2	269.6
decafluorobutane, C_4F_{10} 271		241.4
perfluoropropylmethyl ether, $C_3F_7OCF_3$ 279.9		231.9
perfluorodimethoxymethane, $CF_3OCF_2OCF_3$ 263		226.0
octafluoro-2-butene 270		218.1
octafluorofuran, C_4F_8O 272.42		217.6
octafluorocyclobutane, C_4F_8 267		208.1
perfluorotrimethylamine, C_3F_9N 264		204.0
1,1,1,2,3,3,3-heptafluoropropane, R227ea	257.5	200.1
perfluoroisobutane, C_4F_{10} 252.45		196.8
tris(difluoromethyl)amine, $C_3H_3F_6N$ 275		194.6
trifluoromethyl-1,2,2,2-tetrafluoroethyl ether, E227me	263.6	187.9
1,1,1,2,2,3,3,4-octafluorobutane, R338mccq	301	186.4
1,1,1,2,3,3-hexafluoropropane, R236ea	279.7	185.4
difluoromethyl-1,2,2,2-tetrafluoroethyl ether, E236me	296.5	184.9
octafluoropropane, R218	236.4	183.8
2,2,3,3,3-pentafluoropropanol, $C_3F_5H_3O$ 353.15		173.5
perfluorooxetane, C_3F_6O 244.0		171.9
1,1,2,2-tetrafluorodimethylether, E134	279.4	150.8
bis(trifluoromethyl)ether, E116	257	150.3
hexafluoropropylene 243.5		148.2
pentafluorodimethylether, E125	235	147.1
pentafluoroethane, R125	225.2	134.2
1,1,2,2-tetrafluoroethane, R134	250	133.3
hexafluoroethane, R116	194.95	130.6
1,1,1,2-tetrafluoroethane, R134a	247	129.4
bromotrifluoromethane, R13B1	215.26	100.3
water 373.1		76.0

$$\Delta H_g = \int_{T_b}^{1400K} C_p^{gas} \, dT. \qquad (3)$$

We estimated the gas phase heat capacities using the method of Rihani and Doraiswamy [30], which is based on structural contributions. For agents with boiling points below 298 K, the total heat involved is just the contribution ΔH_v. Table 5 summarizes the results of this computation, with compounds listed in order of their ΔH^{Total} expressed on a molar basis (kJ/mol). Table 6 gives the same quantity, but expressed on a mass basis (kJ/kg). We have also included some common agents such as nitrogen, argon, water and carbon dioxide for comparison.

The results of Table 5 indicate that the top candidates for streaming agents (defined here as having boiling points higher than 293 K), on a molar basis, are perfluoro-2-butyltetrahydrofuran, methoxy-nonafluorobutane, heptafluoropropyl-1,2,2,2-tetrafluoroethyl ether, and 1,1,1,2,2,3,3,4-octafluorobutane.

Table 5. Results of Search for Compounds with High ΔH^{Total} (Molar Basis)

Compound	T_b(K)	ΔH^{Total} (kJ/mol)
perfluoro-2-butyltetrahydrofuran, $C_8F_{16}O$ 375.2		608.4
heptafluoropropyl-1,2,2,2-tetrafluoroethyl ether	315.2	421.9
methoxy-nonafluorobutane, HFE7100	334	407.6
perfluoropropyltrifluoromethyl ether, $CF_3OCF_2CF_2CF_3$ 279.9		336.3
perfluoroisobutane, C_4F_{10} 252.5		321.2
1,1,1,2,2,3,3,4-octafluorobutane, R338mccq	301	321.0
decafluorobutane, C_4F_{10} 271		318.5
octafluorocyclobutane, C_4F_8 267		303.2
perfluorotrimethylamine, C_3F_9N 264		293.8
lactic acid, $C_3H_6O_3$ 455		292.9
perfluorodimethoxymethane, $CF_3OCF_2OCF_3$ 263		286.8
octafluorofuran, C_4F_8O 272.4		289.8
2,2,3,3,3-pentafluoropropanol, $C_3F_5H_3O$ 353.15		280.8
octafluoro-2-butene 270		261.9
tris(difluoromethyl)amine, $C_3H_3F_6N$ 275		256.5
trifluoromethyl-1,2,2,2-tetrafluoroethyl ether, E227me	263.6	255.3
perflurooxetane, C_3F_6O 244.0		252.9
octafluoropropane, R218	236.4	250.8
difluoromethyl-1,2,2,2-tetrafluoroethyl ether, E236me	296.5	239.4
1,1,1,2,3,3,3-heptafluoropropane, R227ea	257.5	236.5
1,1,1,2,3,3-hexafluoropropane, R236ea	279.7	222.2
bis(trifluoromethyl)ether, E116	257	196.2
hexafluoropropylene 243.5		192.9
pentafluorodimethylether, E125	235	183.8
hexafluoroethane, R116	194.95	181.1
1,1,2,2-tetrafluorodimethylether, E134	279.4	170.2
pentafluoroethane, R125	225.2	166.8
1,1,1,2-tetrafluoroethane, R134a	247	156.8
1,1,2,2-tetrafluoroethane, R134	250	152.8
bromotrifluoromethane, R13B1	215.26	110.9
water	373.1	87.5
carbon dioxide	N/a	56.2
nitrogen	77.4	34.7
argon	87.3	22.9

Table 6 indicates that methoxy-nonafluorobutane (HFE7100) is high on the list when expressed on a mass basis. Methoxy-nonafluorobutane is available commercially [23], the others are not available commercially but have been synthesized and used for research [24,25,26]. The top total-flooding candidates (defined as having boiling points lower than 293 K) are perfluoropropyltrifluromethyl ether, perfluoroisobutane, decafluorobutane, and octafluorocyclobutane. Of these, decafluorobutane and octafluorocyclobutane are available commercially. [31] All of these compounds contain only C, H, F, and O.

As noted above, HFE7100 is ranked near the top of the compounds ordered in terms of their predicted ability to extract heat upon heating from room temperature to a typical combustion temperature. This

Table 6. Results of Search for Compounds with High ΔH^{Total} (Mass Basis)

Compound	$T_b(K)$	ΔH^{total} (kJ/kg)
water	373.1	4855.7
lactic acid, $C_3H_6O_3$	455	3251.6
2,2,3,3,3-pentafluoropropanol, $C_3F_5H_3O$	353.15	1872.0
methoxy-nonafluorobutane, HFE7100	334	1630.4
1,1,1,2,2,3,3,4-octafluorobutane, R338mccq	301	1588.7
1,1,1,2-tetrafluoroethane, R134a	247	1536.8
tris(difluoromethyl)amine, $C_3H_3F_6N$	275	1535.5
perfluorooxetane, C_3F_6O	244.0	1523.2
octafluorocyclobutane, C_4F_8	267	1515.8
1,1,2,2-tetrafluoroethane, R134	250	1497.6
heptafluoropropyl-1,2,2,2-tetrafluoroethyl ether	315.2	1475.0
perfluoro-2-butyltetrahydrofuran, $C_8F_{16}O$	375.2	1462.3
1,1,1,2,3,3-hexafluoropropane, R236ea	279.7	1461.4
1,1,2,2-tetrafluorodimethylether, E134	279.4	1442.0
difluoromethyl-1,2,2,2-tetrafluoroethyl ether, E236me	296.5	1424.7
1,1,1,2,3,3,3-heptafluoropropane, R227ea	257.5	1390.9
pentafluoroethane, R125	225.2	1389.8
trifluoromethyl-1,2,2,2-tetrafluoroethyl ether, E227me	263.6	1372.4
pentafluorodimethylether, E125	235	1351.3
perfluoroisobutane, C_4F_{10}	252.5	1349.6
octafluorofuran, C_4F_8O	272.4	1341.5
decafluorobutane, C_4F_{10}	271	1338.2
octafluoropropane, R218	236.4	1333.8
perfluorotrimethylamine, C_3F_9N	264	1329.3
perfluoropropyltrifluoromethyl ether, $CF_3OCF_2CF_2CF_3$	279.9	1323.9
hexafluoroethane, R116	194.95	1312.2
octafluoro-2-butene	270	1309.2
perfluorodimethoxymethane, $CF_3OCF_2OCF_3$	263	1303.5
hexafluoropropylene	243.5	1285.8
carbon dioxide, CO_2	n/a	1276
bis(trifluoromethyl)ether, E116	257	1273.9
nitrogen, N_2	77.4	1238.3
bromotrifluoromethane, R13B1	215.26	744.7
argon	87.3	573.2

particular compound was developed as a replacement for ozone-depleting chemicals used in a variety of applications. It has been estimated to have a zero ozone depletion potential, a relatively low global warming potential, favorable toxicological properties, and good materials compatibility properties. [23] It has been approved under the Significant New Alternatives Policy (SNAP) of the United States Environmental Protection Agency. There is an existing patent dealing with use of HFE7100 as a fire-extinguishing agent. [32] Based on these favorable properties, it was decided to experimentally characterize the extinguishing capability of HFE7100.

The data included in Table 6 indicate that lactic acid is predicted to be highly effective as a thermal agent. Clearly, this compound is not a potential candidate as a fire suppression agent due to its high boiling and

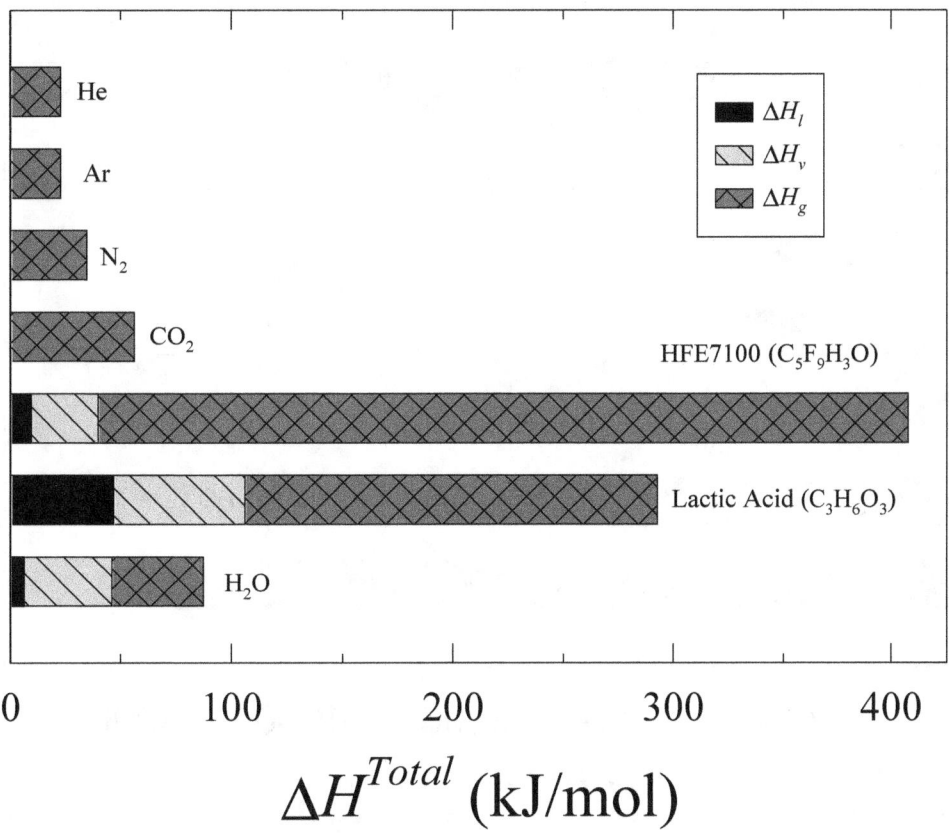

Figure 1. The relative contributions of liquid heating, vaporization, and gas heating to 1400 K to the total heat absorbed by various thermal agents on a per mole basis are shown as bar graphs.

melting points and caustic nature. It is also an organic compound and may be combustible. However, based solely on its ability to extract heat, it is predicted to be nearly twice as effective on a mass basis as any of the other potential thermal agents considered, except water. Its ability to extract heat is predicted to be roughly 2/3 of that of water on a mass basis and 3.5 times greater on a molar basis. Since lactic acid is soluble in water, it was deemed worthwhile to investigate the suppression characteristics of mixtures of these two polar liquids. As described below, the extinguishing effectiveness of water/lactic acid mixtures was experimentally investigated.

In order to better understand the effectiveness of thermal agents, experimental and modeling investigations of flame extinguishment by the widely recognized and investigated gaseous thermal agents nitrogen, carbon dioxide, argon, helium, and water were also carried out. It is of interest to compare the relative contributions of the various heat extraction mechanisms included in Eq. (1). Figure 1 shows a bar plot of the three heat extraction processes for the four gaseous agents, the two liquid agents identified for further investigation, and water in terms of heat extraction per mole of agent. On a molar basis HFE 7100 is predicted to be the most effective, with lactic acid a close second. The contribution of liquid processes to the total heat extraction is relatively small for HFE7100. Liquid water is predicted to be the next most effective due primarily to the relatively high heat absorbed during evaporation. The four gaseous agents are predicted to be relatively ineffective and are ordered $CO_2 > N_2 > Ar = He$.

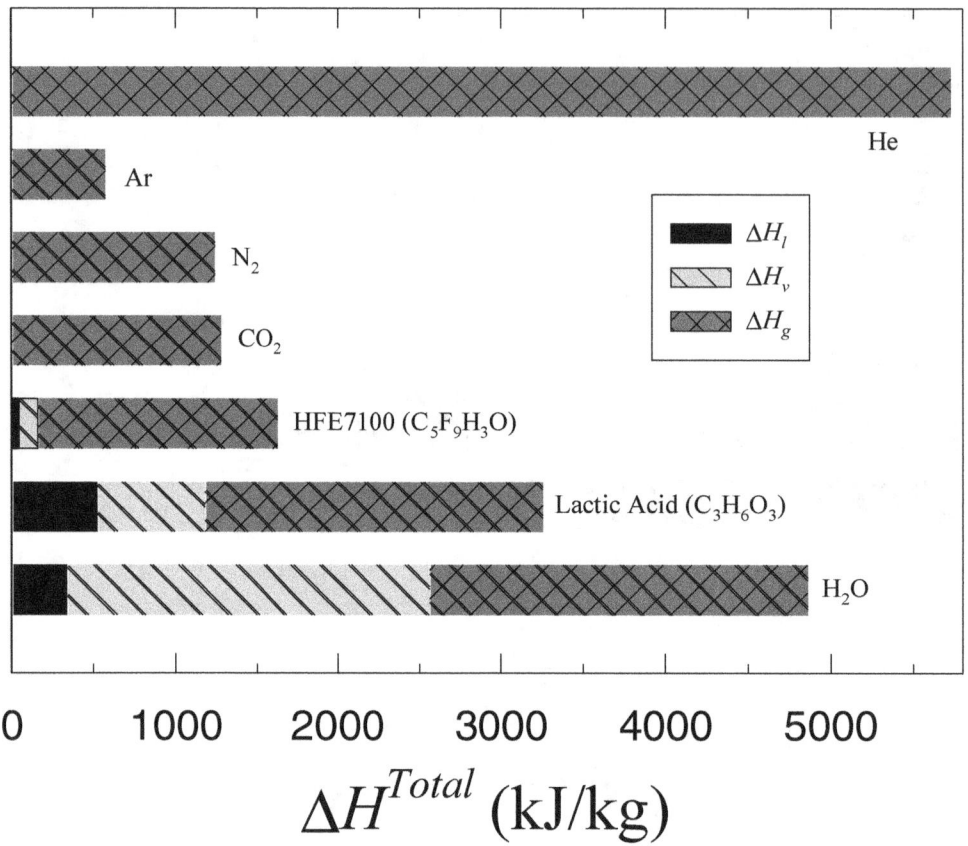

Figure 2. The relative contributions of liquid heating, vaporization, and gas heating to 1400 K to the total heat absorbed by various thermal agents on a per mass basis are shown as bar graphs.

The results included in Figure 1 have been replotted in Figure 2 on a per mass basis. The plot now has a very different appearance. The most effective agent in these terms is helium followed by water and lactic acid. On this plot the role of heat extraction by heating and vaporization of the liquid agents is clear. It can be seen that for water over half of the heat extraction arises from the liquid processes. For lactic acid the contribution of liquid vaporization to the total heat extraction is significant, but greatly reduced as compared to water. The large differences between the HFE7100 and lactic acid and water are due to hydrogen bonding that exists in water and lactic acid. This bonding significantly increases the amount of heat energy required to vaporize liquids when it is present.

Note that nitrogen and carbon dioxide are predicted to extract roughly the same amount of heat per kg, while the amount extracted by gaseous HFE7100 is only slightly higher. In fact, it has been found that for a number of gaseous species expected to act primarily as thermal agents the mass required to extinguish a fire is roughly constant. [33] However, it is clear from the results shown in Figure 2 that this criterion should only be used to provide guidance and that it is not generally applicable to a wide range of molecular species.

Table 7. Some Properties of Lactic Acid Taken From Reference [9]

Property	
Molecular weight (kg/kmol)	90.097
Normal boiling point (K)	490.0
Critical temperature (K)	675.0
Latent heat of vaporization at normal boiling point (J/kmol)	5.92×10^7
Standard heat of combustion (J/kmol)	-1.228×10^9
Flash point (K)	410.0
Lower flammability limits (vol %)	3.1
Upper flammability limits (vol %)	18.0

B. Experimental Testing of Lactic Acid and HFE7100

Two chemicals—lactic acid and HFE7100—were identified in the last section as being particularly interesting as potential highly effective thermal agents. In this section the effectiveness of these agents are tested using experimental screens developed as part of the NGP.

1. Lactic Acid

Table 7 summarizes some of the thermophysical properties of lactic acid. Despite its expected ability to withdraw significant levels of heat from a flame zone, lactic acid is flammable. There will be competitive effects between the heat release and heat extraction in the suppression processes. One approach for mitigating the effect of the heat release is to mix lactic acid with a good thermal agent like water, which is miscible with lactic acid. The mixing of lactic acid with water may also offer potential synergistic effects for fire suppression effectiveness. For these reason, our main focus is the fire suppression performance of various lactic acid and water mixtures delivered to the flame in the form of droplets.

a) Estimation of the latent heat of vaporization for lactic acid/water mixtures

In order to explore the effectiveness of a lactic acid/water mixture as a potential thermal agent, it is necessary to know the latent heat of vaporization for the mixture. The estimation of the latent heat of vaporization for this case is more complicated than for a pure fluid. At constant pressure, the liquid begins to boil once it is heated to the bubble-point temperature for the mixture. At this time, the composition begins to change (unless it is an azeotrope). As the vaporization continues at constant pressure, the more volatile substance (water in this case) concentrates in the vapor phase, while the liquid phase becomes more concentrated in lactic acid. The temperature gradually increases from the bubble point temperature to the dew point temperature. Equilibrium phase diagrams have been constructed for the gaseous and liquid lactic acid/water mixtures. Figure 3 shows a T-x,y diagram (x and y refer to the mole fractions of the liquid and vapor phase, respectively, at a given temperature T) for lactic acid/water mixtures at a constant pressure of 101 kPa (1 atmosphere). The absence of a temperature where the two lines meet indicates that lactic acid and water do not form an azeotropic mixture. The T-x,y diagram was constructed using the Peng-Robinson equation of state with a binary interaction parameter of 0.88 [34].

If one assumes that the entire sample is vaporized, then the integral heat of vaporization at constant pressure can be found using $\Delta H_v = H_d - H_b$, where H_d is the enthalpy of the saturated vapor at the dew

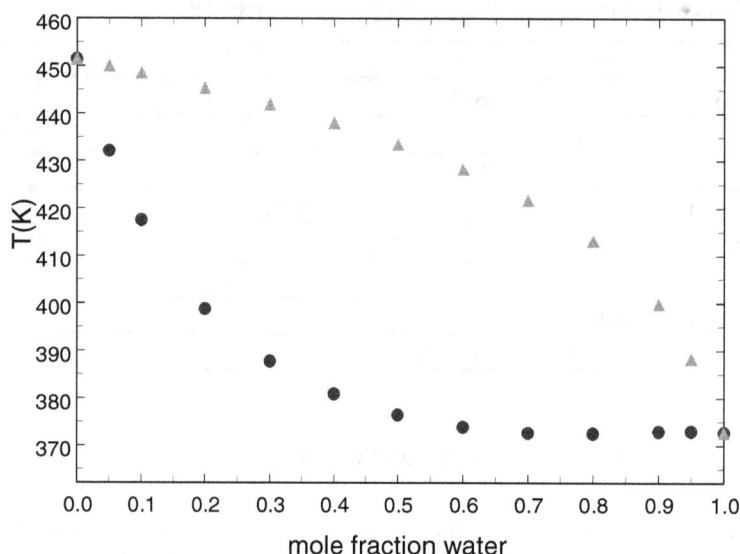

Figure 3. An equilibrium *T-x,y* diagram for lactic acid/water mixtures is shown for a pressure of 101 kPa. The symbols refer to the liquid ( ) and gaseous ( ) phases.

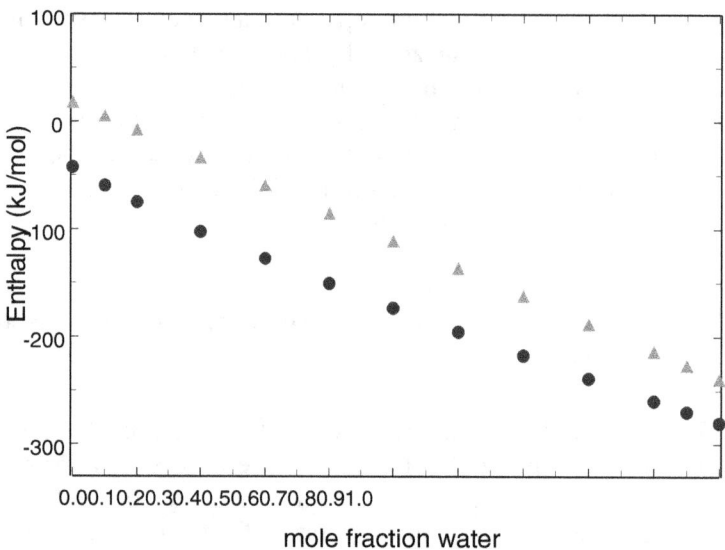

Figure 4. Enthalpies for the saturated vapor at the dew point ( ) and saturated liquid at the bubble point ( ) are plotted as a function of water mole fraction for lactic acid/water mixtures.

point and H_b is the enthalpy of the saturated liquid at the bubble point. Figure 4 shows the enthalpies of the saturated vapor at the dew point and the saturated liquid at the bubble point.

The latent heat of vaporization (kJ/mol) was calculated and fit to a polynomial of the following form,

$$\Delta H_v = 62.35 + 51.71x - 130.1x^2 + 57.92x^3 \ , \qquad (4)$$

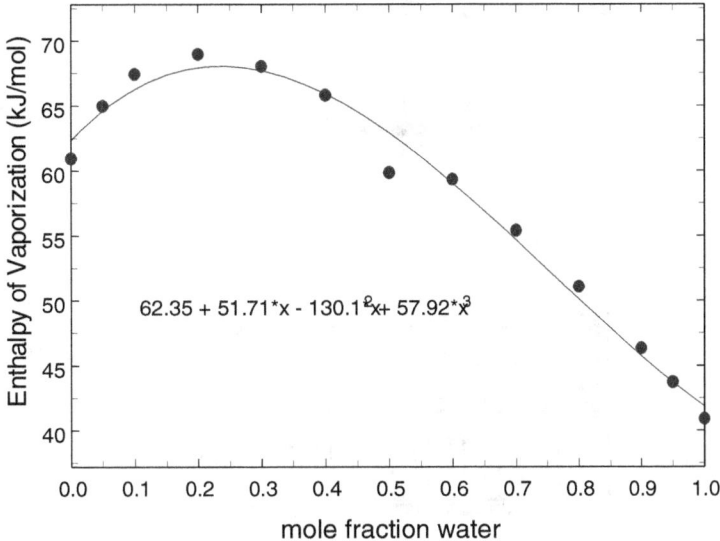

Figure 5. The enthalpy of vaporization for lactic acid/water mixtures for a pressure of 101 kPa is plotted as a function of the water mole fraction.

where x is the mole fraction of water. This relationship is illustrated in Figure 5. Compared to neat water, there is more than a one and a half fold increase in the latent heat of vaporization at a water mole fraction of 0.2.

        b)   Experimental system

The fire suppression effectiveness of lactic acid/water mixtures was examined using the NIST Dispersed Liquid Agent Fire Suppression Screen (DLAFSS) Apparatus. The apparatus was built under the auspices of the NGP, and it is described in a published report. [35] A more detailed description of the apparatus and the experimental procedure is available in a NISTIR [36]. Figure 6 shows a schematic for the DLAFSS.

Briefly described, the apparatus consists of a small vertical wind tunnel, a porous cylindrical burner located at the test section, and a small nebulizer located in the settling chamber of the tunnel. The open circuit wind tunnel incorporates a diffuser followed by a flow conditioning section consisting of honeycomb, screens, and a settling chamber. The settling chamber is connected to the 10 cm × 10 cm square cross section, 20 cm long test section by a contraction. Air is supplied to the tunnel via a frequency-controlled blower.

The burner was located near the base of the test section and was constructed from a 3.18 cm long sintered stainless-steel threaded cup filter with a 1.58 cm outer diameter. Fuel was piped to the inside of the filter and exited through the porous surface. Water cooling was used to limit preheating of the fuel. A thin coat of high-temperature black paint was applied to the end of the filter and one half of the cylindrical surface to limit the fuel flow to a 180 E section of the cylinder face. The burner was aligned in the test section with the open side facing downwards. In this way fuel, was released as a counter flow to the oxidizer flow over an area of 7.9 cm^2. This type of burner configuration is often referred to as a "Tsuji burner". The fuel for the burner was propane with a volume flow rate of 2 L/min, corresponding to a nominal velocity of 4.2 cm/s.

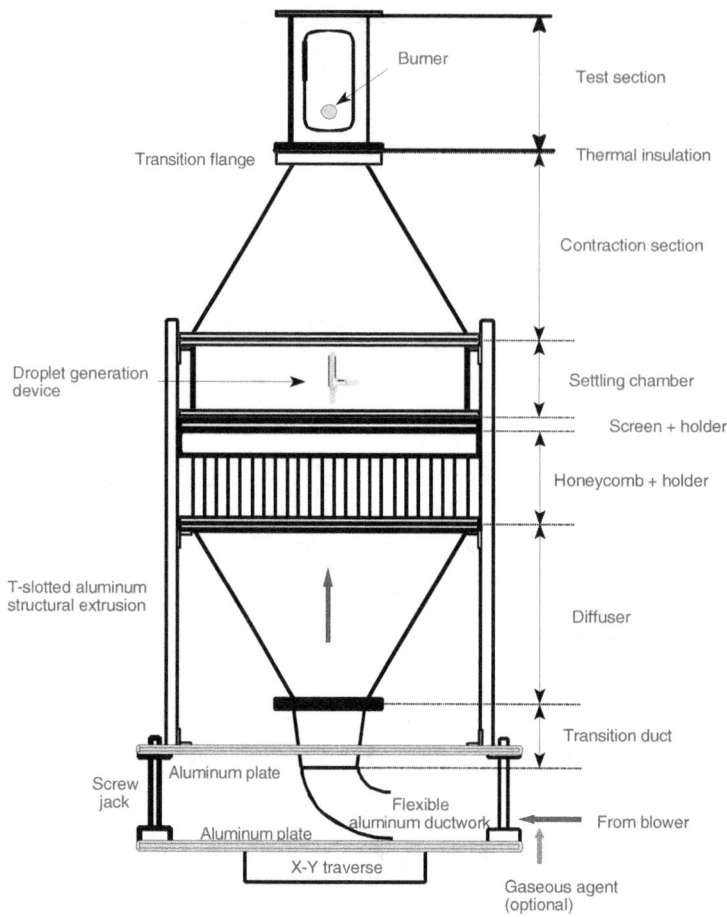

Figure 6. A schematic of the dispersed liquid agent fire suppression screen (DLAFSS) apparatus is shown.

The nebulizer, located in the settling chamber, is used to generate a small poly-dispersed spray of liquid droplets. The droplet Sauter mean diameters (measured using a phase Doppler interferometer) at the burner location without the flame, varied between 25 μm and 35 μm under the test conditions. The experiments were conducted by increasing the airflow through the tunnel with a fixed liquid delivery rate to the nebulizer. The droplets were entrained by the airflow and were transported to the test section. The air velocity at which the flame at the forward stagnation region of the burner was blown-off was used as an indicator for comparing fire suppression efficiencies of various lactic acid/water mixtures.

c) Experimental results for lactic acid/water mixture fire suppression effectiveness

Figure 7 shows the test results for three lactic acid/water mixtures. These mixtures were made by diluting a commercial stock solution of lactic acid in water with a lactic acid mass fraction of 88.5 %. Dilutions of the stock solution with water by one-third, a half, and two-thirds yielded lactic acid mass fractions of 22.1 %, 44.3 %, and 59 %, respectively. The undiluted stock solution was not tested because it was so viscous it caused the syringe pump, used to deliver the liquid to the nebulizer, to stall. Each data point in

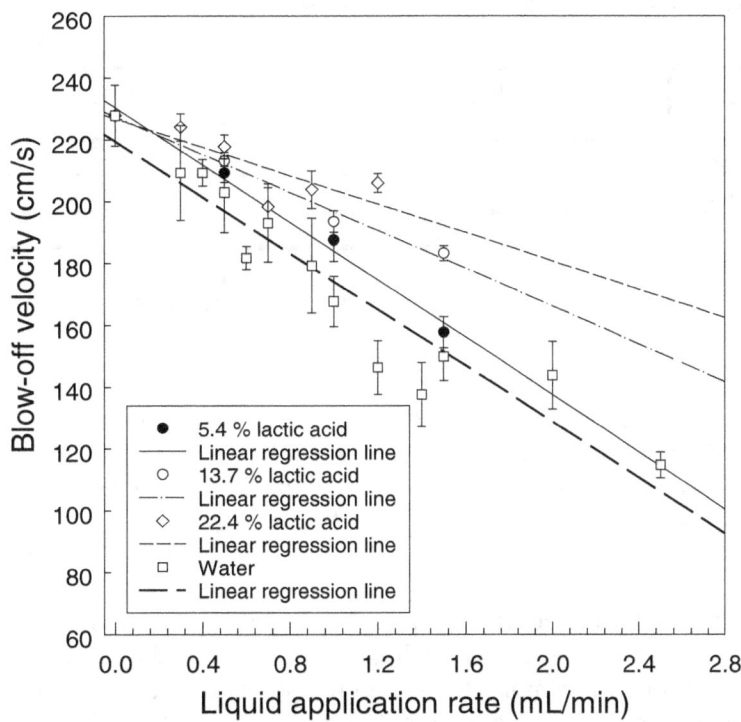

Figure 7. Velocities of the air at extinguishment (indicated by blow off of the opposed-jet diffusion flame) are plotted as a function of liquid application rate for the indicated lactic acid/water mixtures. Error bars represent one standard deviation.

Figure 7 represents an average of 5 or more runs. The error bars represent one standard deviation. The blow-off velocity decreases with increasing liquid application rate. In other words, as more of the acid mixture is delivered to the flame, less airflow is required to induce extinction. Equivalently, the opposed-flow diffusion flame is blown off at lower strain rates.

The slopes of the lines in Figure 7 are a measure for the effectiveness of a liquid in suppressing a flame. The steeper the slope the more effective is the liquid. It can be seen that the three lactic acid/water mixtures show no improvement of fire suppression effectiveness over neat water. In fact, the fire suppression effectiveness decreases as the lactic acid mass fraction increases. Given that no chemical effects are to be expected for fire suppression with lactic acid, it can be concluded that the increased heat extraction due to lactic acid is being overwhelmed by heat release due to the combustion of this organic acid.

Since there are many liquid delivery rates that one can use in the screening procedure, a reference delivery rate is needed to compare and interpret the fire suppression effectiveness of various liquid agents in a consistent way. The following protocol, which is based on the conditions commensurate with cup-burner results for nitrogen, is used.

The average propane cup burner value for nitrogen is 32 % (mass fraction) [33]. An examination of the nitrogen results indicates that the nitrogen mass fraction at blow-off equivalent to the cup burner value corresponds to a reference blow-off velocity of ~ 30 cm/s. [36] Similar reference blow-off velocities were obtained when the cup-burner results for other gases (argon, helium) were used. [36] For this reference velocity, there are some operational issues associated with the burner and the nebulizer [36].

Table 8. Calculated Equivalent Mass and Mole Fraction Values for the Reference Blow-Off Velocity

Agent	Equivalent mass percent (%)	Equivalent mole percent (%)
Water 2.3 3.7		
22.1 % lactic acid	2.5	2.1[§]
44.3 % lactic acid	3.8	3.9[§]
59 % lactic acid	5.1	6.7[§]
HFE7100 11.7 1.5		

[§]Mole-fraction averaged molecular weight of lactic acid /water mixture is used in the calculation.

Therefore, in order to compare the results obtained from the cylindrical burner to conditions commensurate with cup-burner results, extrapolation to the reference velocity is required. Table 8 compares the results at the reference velocity for water and the three lactic acid/water mixtures in both mass and mole fraction terms. The result for HFE 7100 (see below) is also included in Table 8 for comparison purposes.

The results in Table 8 provide further confirmation of the conclusion above that the heat release associated with combustion of lactic acid overwhelms the heat extracted by heating the compound. The amount of lactic acid/water mixture in molar terms required to cause extinction for the reference condition actually increases with increasing mole fraction of lactic acid.

2. HFE7100

HFE7100 has been identified as a particularly promising thermal agent for halon replacement based both on its predicted ability to extract heat and its commercial availability. Two screening tests for extinguishing agents developed as part of the NGP have been used to characterize the effectiveness of HFE7100.

a) Testing of HFE7100 using the Dispersed Liquid Agent Fire Suppression Screen facility

The relative effectiveness of HFE7100 was tested in the DLAFSS using the same experimental procedures described above for the lactic acid/water mixtures. Figure 8 shows a plot of the measured blow off velocity versus the liquid application rate for the HFE7100. Extrapolating to the reference velocity yields effective molar and mass fraction extinguishing concentrations of 1.5 % and 11.7 %, respectively. These values have been included in Table 8.

b) Testing of HFE7100 using the Transient Application Recirculating Pool Fire facility

To further evaluate the fire suppression effectiveness of HFE7100, the agent was introduced as a mist into the Transient Application Recirculating Pool Fire (TARPF) facility. The TARPF was developed originally to assess the effectiveness of gaseous agents for suppressing flames attached behind a bluff body in a turbulent air flow. The facility was designed to generate flow conditions typical of those present in aircraft engine nacelles. This facility was selected as a test bed because this flame configuration is believed to be particularly challenging for liquid agents.

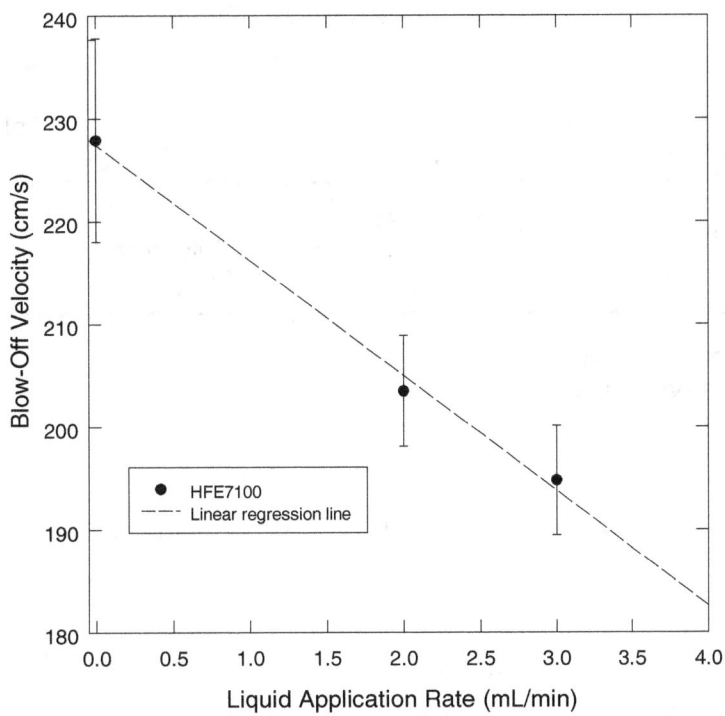

Figure 8. Velocities of the air at extinguishment (indicated by blow off of the opposed-jet diffusion flame) are plotted as a function of liquid application rate for HFE7100. Error bars represent one standard deviation.

Briefly described, the TARPF is a horizontal, open-circuit wind-tunnel with a flat porous burner located behind a baffle or a backward-facing step where a propane flame simulating a pool fire can be stabilized. A *gaseous* agent is transiently or impulsively introduced upstream into the air stream for a short duration. Suppression effectiveness is assessed based on the amount of agent required to extinguish the flame as a function of the application duration and the air flow rate within the tunnel. Detailed descriptions of the TARPF facility have been provided by Grosshandler *et al.* [37,38,39].

For this project the TARPF facility was modified to accommodate the application of a liquid agent. Figure 9 shows a schematic of the modified TARPF used to assess the fire suppression effectiveness of HFE7100. The liquid dispensing system consisted of a liquid reservoir, a compressed air cylinder, two computer-controlled solenoids, and a mist-jet hollow-cone nozzle. The nozzle was positioned 65.5 cm upstream of the porous burner in the horizontal center and at a height corresponding to the midway point between the top of the ramp and the top of the wind tunnel. According to the manufacturer's specifications, the nozzle has a nominal spray angle of 70° when operated at 1.03 MPa with a flow rate of 1.9 mL/s. In order to generate a mist for a fixed duration, the computer activated the solenoid connected to the reservoir. At the end of the discharge, this solenoid was deactivated to terminate the flow to the nozzle, and the second solenoid leading to drain was simultaneously activated to prevent any residual flow to and/or dripping from the nozzle.

For these tests the airflow in the tunnel was fixed at 6.67×10^{-3} m^3/s. The porous burner was operated with a propane flow rate of 33 mL/s. The mist nozzle had a flow rate of 1.9 mL/s at 1.03 MPa. The mist discharge duration was varied from 1 s to 10 s. The hot surface used to investigate re-ignition was not activated in this test sequence. The only heating of this surface came from the impinging flame. Based

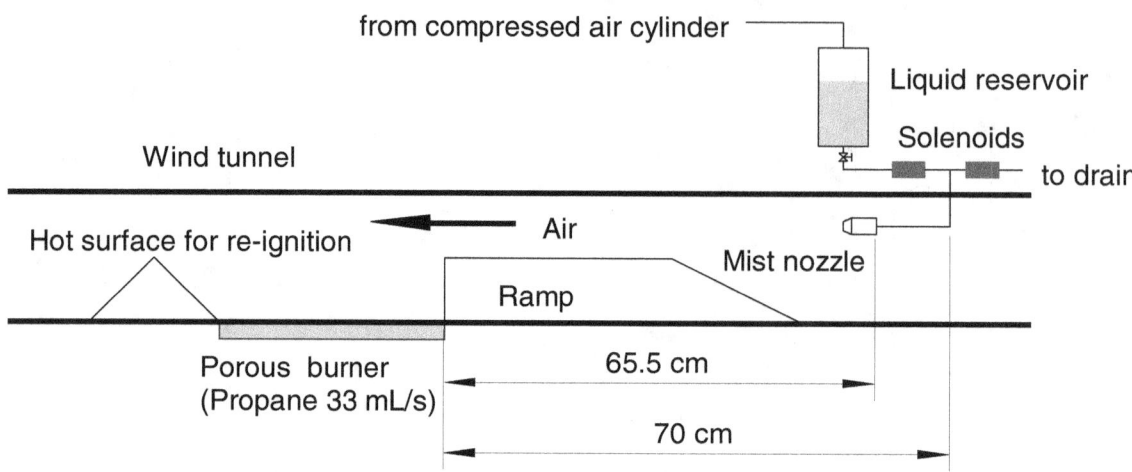

Figure 9. A schematic is shown for the TARPF facility [37,38] as modified for the investigation of the effectiveness of liquid fire suppression agents.

on these conditions, the mass flow rate of the liquid agent, $\dot{m}_{agent}$, can be calculated using the liquid density of HFE7100 (1520 kg/m³ at 25 °C). The mass fraction of the liquid agent in the air stream, Y_{agent}, is then given by

$$Y_{agent} = \frac{\dot{m}_{agent}}{\dot{m}_{agent} + \dot{m}_{air}}, \qquad (5)$$

where $\dot{m}_{air}$ is the calculated mass flow rate of air. Note that in deriving the equation, it is implicitly assumed that the mist droplets are homogeneously dispersed in the carrier phase (air). For the test conditions, the mass fraction of agent was calculated to be 0.30.

For these operating conditions, it was found that the pool fire stabilized behind the backward-facing step could not be *completely* suppressed by the HFE7100 mist. Observations concerning the suppression process can be summarized as follows. Before the mist application, the flame was luminous and yellow. During the mist application, the yellow flame was nearly extinguished with pockets of persistent, blue flames anchored along the flame attachment points at the outer edges of the bluff body. The duration of the small blue flames coincided with the mist application. Once the mist application was over, the flames resumed their original burning intensity and yellow luminosity.

It was somewhat surprising to find that it was not possible to extinguish the propane flames for these test conditions. The nominal effective mass fraction of the HFE7100 for the current experiments is more than twice as high as that required to extinguish a propane flame in the DLAFSS (11.7 %). A similar behavior was observed when water was tested. [39] Possible explanations for these observations are discussed in Section V.A.

III. DETAILED CHEMICAL-KINETIC MODELING AND EXPERIMENTAL INVESTIGATIONS OF THEMAL AGENT FLAME EXTINGUISHMENT

A. Introduction

The extinction of flames by the addition of thermal agents has been the focus of considerable research effort for decades. Much of this effort has been guided by the intuitively reasonable assumption that the addition of nonreactive gases to either the fuel or oxidizer side of a diffusion flame or the fuel/oxidizer mixture for a premixed flame results in extinguishment when the added molecules absorb sufficient heat to lower the temperature in the flame zone to a point at which the combustion reactions are no longer fast enough to generate sufficient heat to overcome local heat losses. For the case of premixed flame propagation, this concept was discussed at least as early as 1925. [40] There has also been considerable work on diffusion flames that are of more direct interest to the current investigation.

Some of the earliest work for diffusion flames was associated with the development of the concept of the limiting oxygen index (LOI) that is often used to characterize the flammability of a fuel. The LOI is defined to be the minimum concentration of oxygen in a mixture of air and nitrogen that is just sufficient to sustain combustion. Simmons and Wolfhard recorded LOI for a wide range of gaseous and liquid fuels using a slow flow of fuel from a hemispherical porous burner into an opposed flow of air diluted with nitrogen. [41] These authors also introduced the concept of a limit flame temperature, which they defined as the adiabatic flame temperature calculated for a stoichiometric mixture of fuel and air at the extinguishing condition.

Simmons and Wolfhard found that the limit flame temperature varied with fuel. [41] For the C1 to C10 normal alkanes the range was 1636 K to 1791 K. Interestingly, the values for methane and the fuels having more than five carbons were all within 20 K of 1791 K. Starting with pentane, the limit flame temperatures fell rapidly as the number of carbons in the fuel was reduced, reaching a minimun value at ethane, before increasing sharply once again for methane. One interpretation of these results is that the smaller multi carbon hydrocarbons, particularly ethane, are more flammable than methane and their longer chain analogs. In 1979 Maĉek considered these results and provided a convincing argument that, in general, the flammability of the smaller straight-chain hydrocarbons, including methane, should increase with decreasing carbon number. [42] This conclusion was based on the behavior of premixed flames and consideration of LOIs for the corresponding normal alcohols. On this basis, Maĉek concluded that when methane is burned as a diffusion flame it has an unexpectedly low flammability.

In the early 1970s the concept of a limit flame temperature was applied in a slightly different way. Huggett [43] and McHale [44] reported that an atmosphere would not support combustion when the available thermal capacity per mole of available oxygen was on the order of or greater than 167 J/K to 209 J/K. Since the heat release per mole of oxygen consumed is roughly constant for most fuels, this finding is equivalent to assuming a common limit temperature.

Tucker et al. explicitly combined the ideas of a constant heat release per mole of oxygen consumed and a limit flame temperature to develop a model for the extinction of diffusion flames by thermal agents. [45] They made the following assumptions in their model:

1. Burke-Shumann diffusion flame model is applicable.
2. Chemical kinetics are identical for premixed and diffusion flames
3. Lower limit of flammability is a thermal limit with a well defined limit temperature
4. In absence of chemical effects, additives do not change the limit temperature.

Experimental extinguishment measurements were made in an apparatus similar to the cup burner [46,47] used for investigating flame extinguishment. Results for four agents, nitrogen, carbon dioxide, helium, and carbon tetrafluoride, which were expected to act as thermal agents, as well as the chemical agent halon 1301, were reported. Three fuels, n-heptane, methanol, and LPG, were used. Oxidizer mixtures having variable amounts of oxygen were prepared. The extinguishing concentrations for each of the thermal agents displayed the dependence on oxygen concentration predicted by the simple model. By assuming a single limit temperature of 1600 K, it was possible to estimate the magnitude of the enthalpy change necessary for extinguishment by each of the agents. These estimates were then compared with actual values calculated from the known heat capacities. It was found that the two results were close, but that the deduced values for carbon dioxide and helium were 11 % and 40 % to 50 % higher, respectively, than the values obtained by direct integration. The authors attributed these differences to chemical effect for the carbon dioxide and the high thermal diffusivity for the helium.

Ishizuka and Tsuji investigated the extinguishment of methane flames burning in mixtures of oxygen with either added nitrogen, argon, or helium. [48] Their flames were stabilized on a porous cylindrical burner in a counterflow of oxidizer. The equivalent of LOIs were reported for each of their special "airs". Thermocouples were used to record flame temperatures just prior to extinguishment. For the methane flames, values ranged from 1443 K to 1623 K for the three inert agents.

Sheinson et al. also considered the effects of thermal agents on flames as part of a larger effort to understand the role of chemical agents on flame extinguishment. [49] They based their analysis on the amount of enthalpy change associated with heating nitrogen and an agent gas from room temperature to 1600 K. The upper value is the assumed maximum flame temperature below which a diffusion flame can no longer be sustained. It is based upon an estimate provided by Roberts and Quince [50] who, in turn, cited an estimate provided by Dodding et al. [51]. No additional references for this choice are cited by Dodding et al. Sheinson et al. recorded cup burner extinguishing concentrations for a number of thermal agents and concluded that the mixture heating of enthalpy was on the order of 250 kJ per mole of oxygen consumed. Consideration of heating effects then allowed the contribution of chemical effects to be estimated.

Zegers et al. have recently performed an extensive test of the approach outlined above. [52] These authors have reported extinction measurements for methane and propane counterflow diffusion flames and cup burner extinguishment measurements for heptane and methanol for a large number of fluorinated hydrocarbons, halon 1301, carbon tetrafluoride, sulfur hexafluoride, and nitrogen. The last three are expected to act primarily as thermal agents. The measurements were compared with concentrations calculated based on the amount of an agent required to absorb an empirically determined amount of heat on going from 298 K to 1600 K. The amount of heat extraction required varied with fuel and was estimated from experimental results for either CF_4 or SF_6 assuming these agents act only thermally. The results were 267 kJ/mol O_2, 267 kJ/mol O_2, 254 kJ/mol O_2, and 258 kJ/mol O_2 for methanol, heptane, methane, and propane, respectively.

The approach provided close estimates for the heavier thermal agents as well as for a number of the fluorinated compounds where chemical effects were small. Interestingly, it overpredicted the amount of nitrogen required for each fuel. The overestimate was largest for methane.

Several other investigations have been reported in which thermal agents have been used to extinguish cup burner flames. In most cases the fuel was heptane. [33,53,54,55,56] Recently, Ural has reported measurements using a cup burner in which methane and propane were burned. [57] The findings of these studies will be discussed more fully below.

A group of researchers have proposed a very different viewpoint with regard to thermal effects on extinguishment of flames and the role of limit temperature. [58,59,60,61] A major conclusion of their analysis is that the vast majority of flame extinguishing agents, including halons, operate primarily by heat extraction, i.e., interference with the flame chemistry is not a dominant mechanism. An early paper by Larsen considered the effects of inert gas agents on flames. [58] He concluded that the effects of thermal agents could be understood entirely based on their heat capacity. He also concluded that the relatively high thermal conductivity of helium did not play a primary role in its extinguishing capability. In later work Larsen extended his analysis to halons and reached the somewhat surprising conclusion that their extinguishing effectiveness could also be understood based entirely on their heat absorption ability. [59]

Ewing et al. also concluded that the effectiveness of many agents recognized as being chemical in nature could be understood based purely on their ability to extract heat. [60] A major difference from earlier work was that while these authors based their analysis on adiabatic flame temperatures, they explicitly assumed that the required limit temperature depended on the suppressant and type of flame system. For nonreactive agents they presented data that indicated that the limit temperature increased with decreasing molecular weight of the agent. It was also argued that the limit temperature varied depending on the type of agent used. Four classes of materials, each with a separate limit temperature, were defined: A) non-decrepitating solids and liquids (2165 K), B) gases and volatile liquids which are fully dissociated in flames (2015 K), C) gases and liquids which are partially dissociated (2015 K), and D) thermally stable gases (variable).

A follow up paper from Ewing et al. developed the idea of variable adiabatic flame temperature limits for extinguishment in much greater detail. [61] Instead of simply defining limit temperatures for different classes of compounds, an approach for estimating the temperature based on the location where heat extraction occurred was used. In fact, it was hypothesized that each agent and flame system has a unique limit temperature and extinguishing effectiveness that depends on the agent's thermal properties. Heat extraction was viewed as either occurring in unburned reaction gases (< 1400 K) or in the combustion zone. The limit temperature was fixed by the relative sizes of the heat removal in the preheat and combustion zones. By assuming certain compounds form predictable secondary products, the authors were able to estimate the heat absorbed by an agent and its dissociation products and estimate the amount required to extinguish the flame as the amount needed to lower the adiabatic flame temperature to the limit temperature. This approach was able to correlate the findings for a wide variety of agents. Despite, the success of this correlation, it has not been widely employed for predictions of agent effectiveness.

It should be clear from this short review that the current understanding of flame extinguishment is based on integral views of diffusion flames, and models for the process utilize the concept of a limit flame temperature that was originally conceived nearly eighty years ago. Meanwhile, within the last two decades the abilities of researchers to calculate simple flame structures using detailed chemical-kinetic models have been developing rapidly and have now reached the point where it is feasible to study flame extinguishment. There have been a number of investigations that have used detailed chemical-kinetic modeling to study flame inhibition and extinction processes. Most of these have involved the reduction of premixed flame speed due to the introduction of various inhibitors. Examples include the early studies of Westbrook [62,63,64] as well as a number of more recent works. [52,65,66,67,68,69,70] Very recently, inhibited flames studies have been extended to laminar opposed-flow diffusion flames. [69,71,72,73,74]

Despite the large number of detailed chemical-kinetic modeling studies dealing with inhibition and extinction of flames by various agents, we have identified only one study which systematically considered the effects of inert thermal agents on a combustion system. Tuovinen calculated the structure of methane/air premixed flames in a well-stirred reactor with added CO_2, H_2O, N_2, and He. [75] This author found that the extinguishing capability of these compounds was proportional to the heat capacity of the gases. The temperatures at extinguishment using the various agents were within 50 K for a given stoichiometric ratio, φ (a range of $\varphi = 0.7$ to 1.2 was studied).

An investigation of the extinguishment of diffusion flames by thermal agents using detailed chemical-kinetic modeling approaches is described in the following section. During the project it became clear that insufficient experimental results for the two fuels considered, propane and methane, were available to fully validate the modeling results. For this reason, extinguishment experiments using two different burners were carried out concurrently and are also summarized.

B. Introduction to Detailed Chemical-Kinetic Modeling of Diffusion Flames

Combustion is a complicated physical process involving the interaction of fluid flow and chemical reaction accompanied by heat release. It is now recognized that combustion involves a large number of chemical species and reactions. The most important reactions are those involving radical species such as OH, HO_2, and the atomic species O and H. The most dominant reaction (see [70] for a recent discussion) is usually recognized as the H atom chain branching reaction leading to breaking of the oxygen bond, i.e.,

$$H + O_2 \rightarrow OH + O. \qquad (6)$$

Subsequent reactions involving the product free radicals release heat and ultimately maintain the combustion. Reaction (6) is strongly temperature dependent, and extinguishment of a flame is expected to occur when the temperature is lowered to a point where the rate of generation of free radicals becomes so slow that the overall reactions cannot generate sufficient heat release to overcome the heat losses from the reaction zone (primarily by thermal diffusion, radiation, and incomplete combustion).

During the past two decades the understanding of the chemical kinetics involved in combustion has reached the level where realistic detailed mechanisms including large numbers of reactants and reactions can be written for simple combustion systems, and mathematical techniques have been developed for simultaneously solving the large number of differential equations that result. While still involving significant approximations (some of which are discussed below), such modeling has developed to the point where it can be used to gain useful insights into the behavior of practically relevant flames.

A number of different types of combustion systems have been modeled including plug flow reactors, perfectly stirred reactors, premixed flames, and opposed-flow laminar diffusion flames. For studies of fire extinguishment, we have chosen to use an opposed-flow laminar diffusion flame model because, for the vast majority of fires, the fuel and air are initially separated and therefore burn as diffusion flames. Most fires are large enough to be turbulent, so a laminar flame model is not strictly correct. However, the most widely used model for turbulent combustion incorporating detailed chemistry is the laminar flamelet model which treats the combustion as laminar flame sheets that are subjected to the local strain rate field associated with the motions of the fluid. [76,77] Thus, the opposed-flow laminar diffusion flame should be the most appropriate flame system available for detailed chemical-kinetic modeling of fire behavior.

There are three configurations of opposed-flow diffusion flames that have been described in the literature—opposed jet, flow over a porous cylinder, and flow over a porous sphere. For the porous bodies, fuel typically flows from the surface, and the surrounding flow is the oxidizer. The three configurations are related in that a diffusion flame is stabilized in the region near the stagnation point in the flow. Numerous papers in the literature discuss these types of flames. Two excellent reviews have been provided by Tsuji [78] and Dixon-Lewis [79]. The review by Tsuji includes a discussion of the use of such burners to experimentally characterize flame inhibition. Examples of early studies include references [80], [81], [82], [83], and [84]. All three configurations are represented. Recently, opposed-jet experiments have been used to investigate potential replacement agents for halons. [35,36,52,85,86]

The most common of the three configurations used in the recent past is the opposed-jet laminar diffusion flame. This configuration has also been the choice for most of the available modeling investigations of opposed-flow flames. For these reasons, we have chosen this configuration for this modeling investigation.

The review of Dixon-Lewis focuses on detailed chemical-kinetic modeling of opposed-flow diffusion flames. [79] Detailed chemical-kinetic models were initially developed in a series of steps. The earliest were designed to describe reactions between H_2 and O_2. The next step was to include CO with the H_2 and O_2. After these mechanisms were sufficiently developed, they were used as the basis for extended mechanisms describing the oxidation of methane. By the early 1980s researchers had begun to calculate the structure of opposed-flow laminar methane/air diffusion flames. The earliest such calculation we have identified is that of Hahn and Wendt that was published in 1981. [87] This was followed by the work of Dixon-Lewis et al. [88], Smooke et al. [89], Dixon-Lewis et al. [90], Puri et al. [91], and Chelliah et al. [92].

Before discussing the findings of these studies, it is important to introduce and clarify some ideas concerning the parameters used to characterize the flow field and its potential interactions with combustion reactions. Much of what follows is based on a review paper by Law. [93]

The degree of interaction between combustion and a flow field is often treated using the concept of flame stretch. This is particularly true for laminar premixed flames. Flame stretch, K, is defined as

$$K = \frac{1}{A}\frac{dA}{dt}, \qquad (7)$$

where A is the infinitesimal area of an arbitrary flame surface lying within the flame structure, e.g., an isotherm or isopleth, and t represents time. In the general case, flame stretch can be broken into three individual contributions due to aerodynamic strain, flame curvature, and flame motion. For a planar, axisymmetric diffusion flame, however, the latter two effects are absent, and K is simply equal to the strain rate within the planar surface, b, with

$$b = \frac{\partial v}{\partial r}, \qquad (8)$$

where cylindrical coordinates have been introduced with z and r the directions perpendicular and parallel to the flame sheet, respectively, and u and v are the velocities in these directions. For the planar flame u is assumed to be only a function of z. The velocities u and v are dependent on each other through the continuity equation, which for a cylindrical coordinate system is written as

$$\frac{\partial(r\rho u)}{\partial z} + \frac{\partial(r\rho v)}{\partial r} = 0, \qquad (9)$$

with ρ equal to the local density.

Law discusses the effects of flame stretch on flame behavior in terms of the different roles of velocity gradient elements tangent to and normal to the flame surface. He notes that there has been a great deal of confusion in the literature in distinguishing between the roles for each element. He argues that it is the normal element of the axial velocity gradient which determines the residence time within the flame and that the Damkohler number, *Da*, defined as the ratio of the characteristic flow time to a characteristic reaction time, should be based on this velocity gradient element. As we shall see shortly, *Da* is a dominant parameter for describing flame extinction. For now, it is important to note that for planar diffusion flames the most important characteristic flow parameter is the element of the axial velocity gradient in the axial direction,

$$a = \frac{\partial u}{\partial z}. \qquad (10)$$

Returning now to the discussion of early detailed chemical-kinetic investigations of methane/air diffusion flames, the paper by Dixon-Lewis et al. is particularly interesting because it involved an intercomparison of calculated results for an opposed-flow methane/air diffusion flame by five different research groups. [88] Calculated flame structures were compared with the experimental measurements of Tsuji and Yamaoka [94,95] for the configuration of flow over a porous cylinder. All of the researchers used the same model for the velocity field. A similarity transform was used to convert the two-dimensional flow field to a pseudo one-dimensional flow, thus greatly reducing the amount of computation required. The coordinate system employed for these calculations was actually rectilinear, but the following discussion will be in terms of cylindrical coordinates. The opposed flows for ambient locations outside of the reacting boundary layer were assumed to be potential flows generated from point sources located at infinity. One appealing aspect of such a flow is that the characteristic strain rate, usually defined as the radial velocity gradient element in the radial direction,

$$b_p = \frac{\partial v}{\partial r}, \qquad (11)$$

is constant in the isothermal region. The continuity equation allows the following relationship between a_p and b_p to be derived,

$$a_p = \frac{\partial u}{\partial z} = -\frac{1}{2} b_p. \qquad (12)$$

Dixon et al. concluded that the models described the qualitative flame structure quite well, but that the quantitative agreement was not perfect. [88] In particular, they argued that the assumption of potential flow was inadequate to describe the actual flow field. The results from the different investigators within the intercomparison also differed somewhat, and additional study was suggested to identify the sources of these variations.

While the strain rate dictated by the potential flow provided a natural variable to characterize the velocity field, the authors also noted that the scalar dissipation rate at the stoichiometric surface of the flame, χ_s, is more commonly used to describe the effect of strain on diffusion flames in turbulent models. [76,77] The scalar dissipation rate is defined as

$$\chi = 2D \left(\frac{\partial Z}{\partial x} \right)^2, \qquad (13)$$

where *D* is a molecular diffusion coefficient and *Z* is the fuel mixture fraction, a conserved scalar. An expression was provided which allowed χ_s to be calculated from the flow parameters. Most of the calculations were performed for strain rates of 100 s^{-1}, but one set was carried to higher velocities where flame extinction occurred. The calculated value of a_p at extinction was 410 s^{-1}, and the corresponding value of χ_s was 16.7 s^{-1}. [88]

The limitations with regard to the use of potential flow conditions in opposed-flow flame calculations continued to worry early workers in the area. It was recognized that most experimental opposed-jet configurations were designed to generate plug flows either through the use of nozzles or active flow straighteners. In other words, the radial velocities at the burner exits were set to be zero, with both a and b also equal to zero at these locations. Furthermore, the nozzles had finite diameters as opposed to the infinite diameters assumed in the model. In 1988 Kee et al. demonstrated that it was possible to derive a pseudo one-dimensional similarity transform for problems that relaxed the conditions imposed by the assumption of potential flow that required b_p and a_p to be constant throughout the isothermal region of the flow. [96] In their formulation, it is only required that the radial velocity profile vary linearly with radial position for a given axial position, as opposed to having a constant value. Thus b is allowed to vary along the primary flow direction. This approach allows the case of initial plug flows, i.e., $\partial v / \partial r = \partial u / \partial r = 0$ at the burner exits, to be evaluated. The price paid for relaxing the potential flow boundary conditions was that the strain rate element normal to the flame surface, a, was no longer a constant for regions outside of the reacting boundary layer, but now varied with the similarity coordinate. Thus there was no longer a unique strain rate that could be used to characterize the flow field. Kee et al. used the approach to calculate the structure of opposed-flow premixed methane/air flames.

The formulation used by Kee et al. [96] is related to earlier work by Seshadri and Williams [97] who considered flow in a two-dimensional channel. Their results can be used to derive a characteristic global strain rate, a_g, for the pseudo one-dimensional flow, which is written for axisymmetric flow as

$$a_g = \frac{-2u_o}{L}\left[1 + \frac{u_f}{(-u_o)}\sqrt{\frac{\rho_f}{\rho_o}}\right], \quad (14)$$

where L is the separation distance of the two burners, u_f and u_o are the fuel and oxidizer velocities, which have opposite signs for the opposed flow, and ρ_f and ρ_o are the fuel and oxidizer densities. This corresponds to the strain rate that would exist at the stagnation point for opposed flows if the boundary layer were infinitely thin. Flow fields with comparable values of a_g are expected to have similar spatial variations of velocity. It should be mentioned that Eq. (14) is occasionally found in the literature without the factor of 2. The form of Eq. (14) is correct for axisymmetric opposed flows. A good discussion of this point is provided by DeCroix and Roberts. [98]

Chelliah et al. [92] extended the formulation provided by Kee et al. [96] to treat opposed-flow methane/air diffusion flames. Calculations were made for two different boundary conditions corresponding to either potential flow with a characteristic strain rate of a_p for the boundaries located at $x = 0$ and $x = L$ or plug flow conditions with strain rate $a(x = 0, x = L) = 0$ s^{-1} at the burner exits. An important feature of this work was comparison of the calculational results with detailed experimental measurements of the velocity field. The flows were generated by nozzles and were expected to be closely top hat in shape. Therefore, plug flow conditions at the burner exits would be expected to provide better agreement with experiments. This was indeed found to be the case, although small disagreements were still observed.

Recall that plug flow conditions at the burner exits require that a varies with axial location. The global strain rate given by Eq. (14) is one means for quantification. However, the reacting boundary layer is known to have a finite thickness, and the volume expansion associated with the heat release causes a to vary rapidly through the combustion region. As a result, the value of a at the stagnation plane does not actually correspond to a_g. For this reason, the measured value a on the oxidizer side of the flow just outside of the boundary layer, a_o, has been suggested for use as the characteristic strain rate for opposed-flow diffusion flames formed by plug flows. Chelliah et al. found that the values of a_g and a_o differed significantly. [92]

The calculations of Chelliah et al. were carried to high enough velocities to cause extinction of the flame. [92] For plug flow conditions the corresponding values of a_g and a_o at extinction were 610 s^{-1} and 391 s^{-1}, respectively, while the experimental value of a_o was 380 s^{-1}. Values of χ_s at extinction were also determined from the calculations. Interestingly, these values were nearly independent of the exit flow conditions, being roughly 14.7 s^{-1} in both cases. The experimental value was reported as 19.9 s^{-1}.

The existence of the various definitions of characteristic strain rates, i.e., a_p, a_g, and a_o have resulted in some confusion in the literature when different experiments and calculations are considered. Fisher et al. have compared values of a_o measured in a propane/air counterflow diffusion flame with the corresponding oxidizer air flow rates and values of a_g. [52,99] In each case, a linear relationship was found. This suggests that for a given flow configuration, values of a_g and a_o should be related by a constant.

Early calculations involving opposed-flow diffusion flames assumed the combustion occurred adiabatically and did not allow for the possibility of radiative heat loss. Recently, however, detailed chemical-kinetic calculations for methane/air diffusion flames have begun to appear in which the effects of radiation from combustion gases have been included. [100,101,102,103] The study by Maruta et al. considered laminar diffusion flames in the opposed-jet configuration burning nitrogen-diluted methane. [100] Both detailed chemical-kinetic modeling and experiments were reported. The experiments were performed in microgravity so that buoyancy effects on the flame were minimal. Their results showed that at high strain rates flame extinction was due primarily to strain effects. However, as the strain rate was lowered, the percentage of heat loss by radiation increased rapidly, lowering the flame temperature and ultimately extinguishing the flame. As a result of these competing effects, plots of maximum flame temperature versus strain rate showed a maximum, and plots of strain rate at extinction as a function of methane concentration in the fuel were double valued.

For their investigation, Chan et al. used a different modeling approach. [101] Their calculation is based on a steady-state laminar flamelet model which is solved in mixture fraction coordinates. The effects of strain were incorporated indirectly by varying χ_s. Both detailed chemical kinetics and radiation losses were included. These authors considered a methane/air diffusion flame. Their findings are consistent with those of Maruta et al. [100] The effects of radiation were minimal at high strain rates and extinction occurred due to the well known strain-rate mechanism. Extinction was calculated to take place at $\chi_s = 18.4$ s^{-1}, and the maximum flame temperature at extinction was 1773 K (1771 K with radiation losses included). However, as the strain rate was lowered, the percentage of the total heat release rate loss due to radiation increased dramatically. Calculations without radiation indicated a continual increase in maximum temperature with decreasing χ_s, while the inclusion of radiation resulted in maximum temperatures that first increased, but then began to fall for values of $\chi_s < 2.5$ s^{-1}. The falling temperature ultimately led to a second extinction at $\chi_s = 0.029$ s^{-1}.

Gore et al. also found that radiation effects become more important at low strain rates. [102] Their calculations showed that there is a substantial range of strain rate for which the effects of strain rate and radiation losses nearly cancel, resulting in maximum flame temperatures that are roughly independent of strain rate. The recent study of Bai et al., which includes laminar flame calculations for hydrogen, methane, and propane flames, also showed that radiation could extinguish a flame at low strain rate. [103]

C. Modeling Approach

1. CHEMKIN III and OPPDIF

The current investigation seeks to calculate the effects of adding thermal agents to the air side of laminar opposed-flow diffusion flames. There are a number of codes described in the literature for making such calculations. We desired to use a code that was easily accessible to researchers. The code OPPDIF [104] developed by Sandia National Laboratories was chosen. This code is now available commercially from Reaction Design[*] of San Diego, CA. OPPDIF is built on a number of general purpose subroutines, collectively known as CHEMKIN-III [105], which handle many of the tasks associated with the calculation. Data describing the reaction mechanism and thermodynamic and transport properties are either incorporated in OPPDIF or accessed as databases.

OPPDIF solves the psuedo-one-dimensional equations describing an axisymmetric opposed-flow laminar diffusion flame. The equations are those originally reported by Kee et al. [96] for premixed flames to solve problems in which b is constant at the boundaries and were later extended to diffusion flames by Chelliah et al. [92] This includes the case of plug flows, i.e., $b = a = 0$ s^{-1} at the burner exits, which was used for all of the calculations that follow.

2. Detailed Chemical-Kinetic Mechanism for Methane

A detailed chemical-kinetic mechanism is required to model a flame. After reviewing the literature, we chose the widely used methane/air mechanism developed with the support of the Gas Research Institute. The version used was GRI-Mech 1.2. [106] At the time this investigation was initiated there was a later version of the mechanism, GRI-Mech 2.11, in which nitrogen chemistry was added. The earlier version was chosen for use since NOX formation is not of interest for the current investigation. Subsequent to the start of the project, GRI-Mech 3.0 was released. It incorporates new reaction data and updated optimization. A comparison of extinction calculations using the three GRI-Mech versions (described below) suggests that the choice of version has a minimal effect on the calculated results.

GRI-Mech 1.2 consists of 32 chemical species undergoing 177 reactions. One- and two-carbon species are included. Thermodynamic and transport data are provided as separate files. The mechanism was created by starting with appropriate estimates for the rate constants and then optimizing to provide the best agreement with experimental measurements in such systems as premixed flames, shock tubes, and flow reactors. [106] Note that the mechanism was not optimized using diffusion flame results.

3. Detailed Chemical-Kinetic Mechanism for Propane

Detailed reaction mechanisms for propane combustion have not been as extensively developed as for methane, and no widely accepted version comparable to GRI-Mech is available. A mechanism assembled at NIST by Babushok and Tsang was used. [70] This mechanism was created by adding a block of reactions for C3-C4 compounds taken from Marinov et al. [107,108] to a C1 and C2 mechanism previously developed at NIST [67]. It consists of 83 species undergoing 509 reactions. Since this

---

[*]Certain commercial equipment, instruments, or material are identified in this paper in order to adequately specify the experimental procedure. Such identification does not imply recommendation or endorsement by the National Institute of Standards and Technology, nor does it imply that the materials or equipment are necessarily the best available for the purpose.

mechanism has not been published in detail, it is included here as Appendix A. Thermodynamic and transport data are included with the mechanism as separate files.

4. Modeling Details

OPPDIF was mounted on a Silicon Graphics Incorporated Origin 200 workstation. An initial test for a laminar hydrogen/air opposed-flow diffusion flame yielded good agreement with the solution provided in the documentation. By simply replacing the mechanism provided for the hydrogen/air flame with GRI-Mech 1.2, we were able to obtain an initial solution for a methane/air diffusion flame.

Calculations for the propane flame proved to be much more difficult. Considerable effort was required to identify an initial set of input conditions that would result in a converged solution for the diffusion flame. Appropriate conditions were eventually identified by a trial-and-error procedure.

The purpose of our calculations is to determine extinction behavior for the flame as functions of flow conditions and agent concentration and to use this information to better understand the effectiveness of thermal agents for extinction of diffusion flames. Extinction of diffusion flames can be understood in terms of the "S" curve that was originally described by Fendell [109] and Liñán [110]. The name refers to the shape of a plot of maximum flame temperature versus Damkohler number, Da, where Da is the ratio of a characteristic flow mixing time and a characteristic combustion reaction time. The unique shape results from the highly nonlinear temperature dependence of combustion reactions. These authors showed that there were two possible solutions for mixtures of fuel and air—combustion and very slow reaction regimes. These two branches are connected by an unstable third branch that forms the central part of the "S", but is not physically observable. Starting well up on the burning branch, it is found that the maximum flame temperature decreases with decreasing Da. As the Da is decreased still further, the temperature begins to drop faster and is ultimately predicted to curve onto the middle branch and continue to fall off with increasing Da. Such a dependence on Da is physically unrealistic, and, as a result, the real flame system drops to the lower branch at the turning point. This is interpreted as flame extinction. A similar behavior applies to ignition. Starting on the lower branch, as the temperature is increased the reaction rate increases very slowly at first. However, a temperature is eventually reached where the rate begins to increase quickly and ultimately begins to increase with falling Da, passing onto the middle branch. However, due to the instability of this branch, the system actually jumps rapidly up to the burning branch. This is the process of spontaneous ignition.

The goal is to identify the flow conditions just sufficient to cause extinction of a flame formed by fuel and either air or air with added agent. Thus the problem becomes one of identifying the turning point for a plot of temperature versus Da. Either the velocity, appropriate characteristic strain rate, or stoichiometric scalar dissipation rate can be used as the independent variable, and since the mixing times associated with these parameters are inversely related to their values, the curve is reversed from that for a Da plot, with the maximum flame temperature decreasing with increasing velocity, characteristic strain rate, or scalar dissipation rate.

Early calculations of flame extinction simply increased the controlling parameter for burning solutions until further increases resulted in a nonburning solution (e.g., see [88]). Due to the rapid changes in temperature with flow parameters near the extinction point, solutions of opposed-flow diffusion flames can become unstable, and there may be some uncertainty associated with the extinction value. As a result, so-called continuation methods have been implemented which allow calculations to be extended onto the unstable solution branch. [111,112] In this way it is possible to determine the turning point corresponding to extinction unambiguously.

Unfortunately, continuation approaches are not implemented in OPPDIF. For this reason, we chose to employ the direct approach in which higher and higher flow velocities were used until either a nonburning solution was found or the differential solver was unable to find a solution. In order to minimize the associated uncertainties, very small incremental increases in velocity were utilized near the extinction point (the solver is more efficient when an earlier solution for similar conditions is used as the starting point). An arbitrary, but stringent, criterion was adopted which defined the extinction velocity as that for which a burning solution was still obtained, but for which an increase of only 0.01 cm/s in the fuel and air initial velocities either resulted in a failure to find a solution or a nonburning solution. Additional small increases in velocity always resulted in nonburning solutions. As we shall show, plots of maximum flame temperature versus the controlling parameters (e.g., velocity) indicate that the flames were indeed near the extinction point as reflected by large temperature drops for small changes in the control parameter. It is believed that this approach results in small errors for estimates of the calculated extinction condition.

A number of parameters are relevant to OPPDIF calculations. [104] Many of these are summarized in Table 9 for the methane/air and propane/air cases. The general procedure was to find a solution for a given type of diffusion flame by first solving an isothermal case and using the solution as the input for a calculation in which the energy equation was added. Once a solution was available, it could be used as the starting point for calculations in which such parameters as the oxidizer and flow velocities were changed. Only the formulation including the energy equation was solved for the "restarted" cases. The new solution then became the starting point for cases involving further changes in the controlling parameters. OPPDIF allows the grid for the calculations to be adapted to best match the latest solution. This feature was used frequently for these calculations. Normally, between 130 and 150 grids points were used for a final solution for the methane flames, while 95 grid points would be typical for a propane flame. For simplicity, the opposed (i.e., opposite signs for velocity) fuel and oxidizer exit velocity magnitudes were always set equal to each other. The exit temperatures were 300 K.

In general, it was found that calculations for the methane flames required considerably less time and were much more computationally stable than those for propane flames. For these reasons, the largest number of calculations was performed for the methane flames. A number of aspects of flame extinguishment were investigated using surrogate agents. These were done primarily using the methane flame.

The final solutions for a particular problem were saved in binary files. A program named POSTOPP was supplied by Reaction Design [104] which reads these files and generates ASCII files containing such parameters as temperature, density, fuel mixture fraction, axial and radial velocities, and axial and normal strain rate elements a and b as functions of distance from the fuel nozzle ($z = 0$ cm). This program was modified to also output the largest absolute value a on the oxidizer side, a_o, the stoichiometric mixture fraction, and the scalar dissipation rate at the location corresponding to a stoichiometric mixture fraction. The derivatives required in Eq. (13) were calculated by simple forward differencing, and values of D were corrected for variations in the local temperature using the expression recommended by Norton et al. [113]

D. Model Results for Methane and Propane Flames Burning in Air

Figure 10 shows a plot of maximum temperature, T_{max}, versus flow velocity (equal magnitudes for the fuel and oxidizer opposed velocities) calculated for a laminar opposed-flow methane/air diffusion flame. The fuel is 100 % methane, and air is assumed to be composed of 78.1 % N_2, 21.0 % O_2, and 0.9 % Ar. As expected, T_{max} decreases with increasing velocity. Extinction occurs for a velocity of 320.12 cm/s for a maximum flame temperature of 1785 K. A global strain rate, a_g, of 558 s^{-1} is calculated for the extinction condition using Eq. (14). Experimental values in the literature include those of Puri and Seshadri [114] (a_g =of 542 s^{-1}), Chelliah et al. [92] ($a_g = 610$ s^{-1}) and DeCroix and Roberts [98] ($a_g = 590$ s^{-1}) The calculated value is in reasonable agreement with these results.

Table 9. Some Parameters Used in Most OPPDIF Calculations of Methane and Propane Flames

Keyword	Methane	Propane
ENRG	Energy Equation Included	
NOFT	Skip the Fixed Temperature Problem	
RSTR	Restart From Earlier Solution	
GRAD 0.3		0.5
CURV 0.3		0.5
IRET (retirement period)	20	
RTOL (relative convergence for Newton's method)	1.E-3	
ATOL (absolute convergence for Newton's method)	1.E-6	
ATIM (absolute convergence for timestepping)	1.E-6	
RTIM (relative convergence for timestepping)	1.E-3	
XEND (physical length)	2 cm	
TFUE , TOXI (fuel and oxidizer initial temperatures)	300 K	
PRES 1	atmosphere	
AFUE, AOXI (radial gradient at inlet)	0 s^{-1}	
MIX	Mixture-Averaged Formula for Diffusion Velocities	

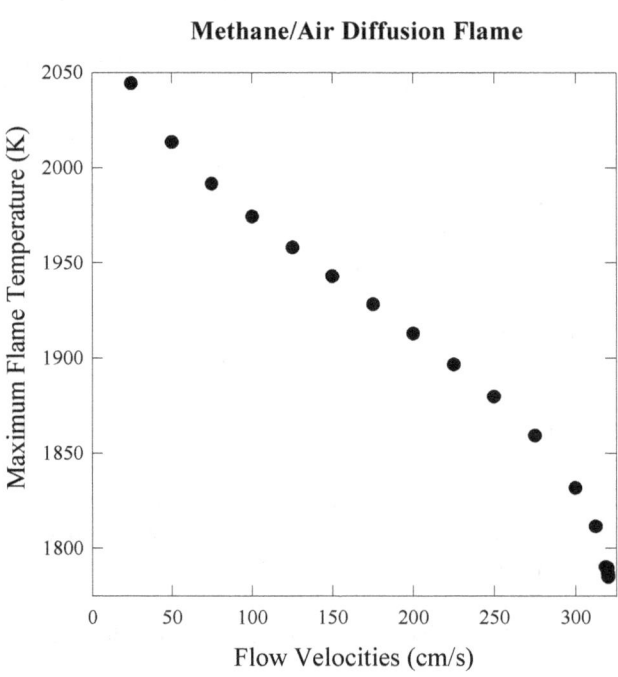

Figure 10. Maximum flame temperatures are plotted as a function of the equal fuel and air velocity magnitudes for a methane/air opposed-flow diffusion flame.

Figure 11 shows two examples, one for relatively low fuel and air velocity magnitudes and one for the methane flame very close to extinction, of the variation of the local characteristic strain rate, a, with position. Several interesting behaviors can be seen. First, as expected, absolute values of a are higher for the case with higher initial velocities. For regions outside of the boundary layer, a values increase or decrease roughly linearly with increasing position, but large intense variations are evident in the vicinity

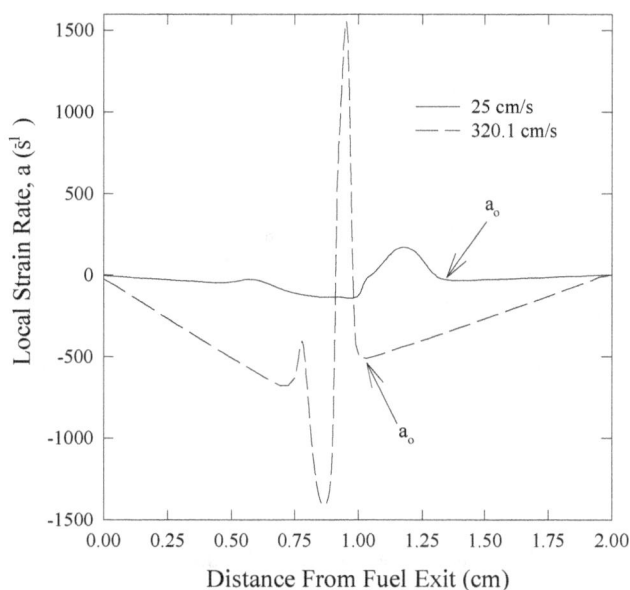

Figure 11. Calculated strain rates, a, are plotted as a function of distance from the fuel exit for low exit velocity magnitudes (25 cm/s for fuel and air) and for a condition very close to extinction (320.1 cm/s).

of the flame front. These large fluctuations within the flame zone result from the volume expansion due to heat release. It is clear that the boundary layer width for the low velocity case is considerably broader than for the higher flow velocity. This is expected since higher absolute velocities should more effectively overcome the effects of molecular diffusion and result in a narrower mixing layer.

The first maximum in the absolute value of a on the oxidizer side, denoted as $|a_o|$, (see Figure 11) can be taken as representative of the strain rate just outside of the boundary layer. Figure 12 shows a plot of maximum flame temperature versus $|a_o|$ for the various flow velocities. As expected, the maximum flame temperature decreases with increasing $|a_o|$. Extinction is calculated to occur when $|a_o|$ is increased to 509 s^{-1}. At least seven experimental measurements of $|a_o|$ at extinction are available in the literature for methane/air diffusion flames (if listed, uncertainties are ± 1 σ): Papas et al., $|a_o| = (405 \pm 25)$ s^{-1} [85], Chelliah et al., $|a_o| = 380$ s^{-1} [92], Du et al., $|a_o| = (380 \pm 20$ s$^{-1})$ [115], Yang and Kennedy, $|a_o| = 340$ s^{-1} [116], Du and Axelbaum, $|a_o| = (375 \pm 10)$ s^{-1} [117], Pellett et al., $|a_o| = 384$ s^{-1} and 396 s^{-1} for two different burners [118], and Zegers et al., $|a_o| = (415 \pm 32)$ s^{-1}. [52] These values are equal within experimental uncertainty. The value calculated here is roughly 30 % higher. Two possible reasons for this are different boundary conditions in the experiments and model and the possible dependence of the results on chemical mechanism. Reasons for the difference are discussed further below.

Figure 13 shows the mixture fraction, i.e., the fraction of the mass of the combustion mixture derived from fuel, plotted as a function of distance from the fuel source for the two same conditions used for Figure 11. The narrowing of the boundary layer due to increased strain rate is clearly evident. The shape of the curve becomes more asymmetric as the velocity decreases. Since values of the mixture fraction are now available, it is possible to calculate the scalar dissipation rate, χ, as a function of x using Eq. (13). Figure 14 shows the results, which demonstrate the strong dependence of χ on the flow velocities. The value of χ at the stoichiometric surface, χ_s, can now be determined by interpolating the results to find the value of χ for a mixture fraction of 0.055, the stoichiometric value. The results are $\chi_s = 1.70$ s^{-1} and $\chi_s = 20.7$ s^{-1} for fuel and air exit velocity magnitudes of 25 cm/s and 320.1 cm/s, respectively.

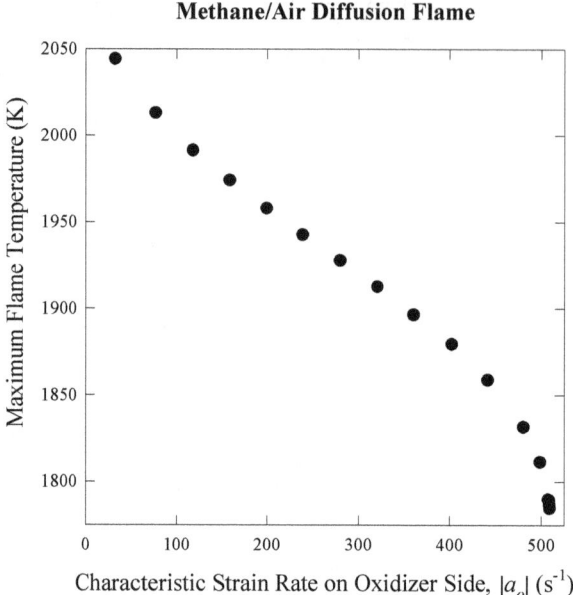

Figure 12. The maximum flame temperature is plotted against a characteristic local strain rate, $|a_o|$, defined to be equal to the highest absolute value of a on the air side observed outside of the boundary layer associated with the methane/air flame.

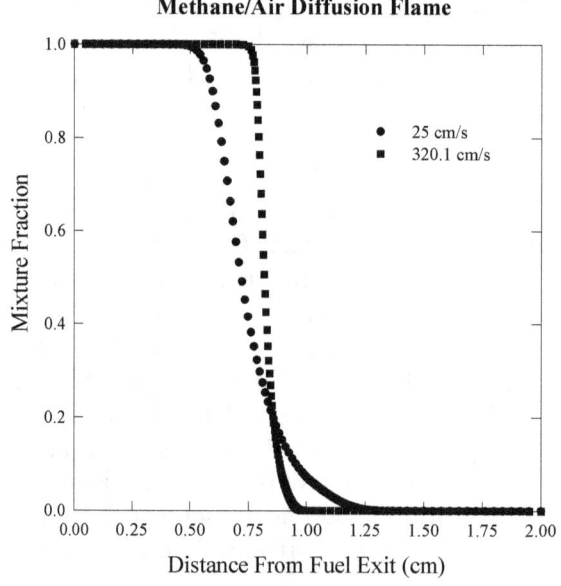

Figure 13. Mixture fraction is plotted as a function of distance from the fuel exit for low exit velocity magnitudes (25 cm/s for fuel and air) and for a condition very close to extinction (320.1 cm/s) for a methane/air diffusion flame.

Maximum calculated flame temperatures are plotted as a function of χ_s for a series of methane/air flames in which the fuel and air velocities were varied in Figure 15. The temperature drops with increasing χ_s, and extinction is calculated to occur for $\chi_s = 20.7$ s^{-1}. No experimental measurements of χ_s were

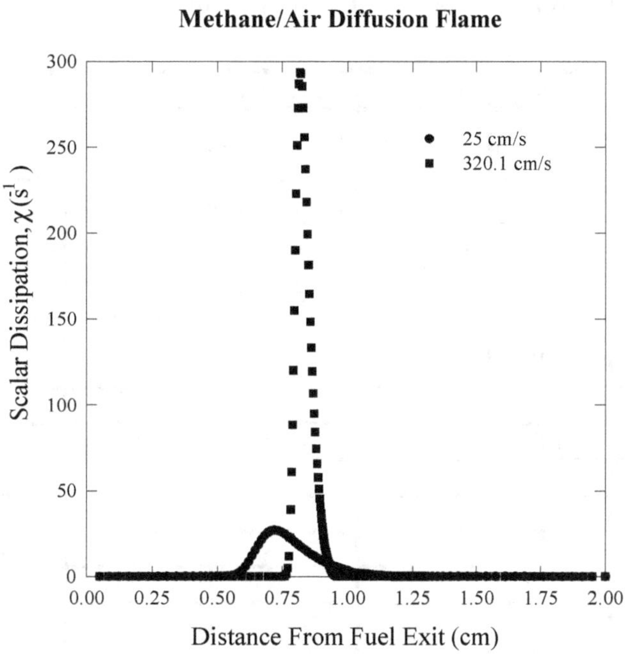

Methane/Air Diffusion Flame

Figure 14. Calculated scalar dissipation rates, χ, are plotted as a function of distance from the fuel exit for low exit velocity magnitudes (25 cm/s for fuel and air) and for a condition very close to extinction (320.1 cm/s) for a methane/air diffusion flame.

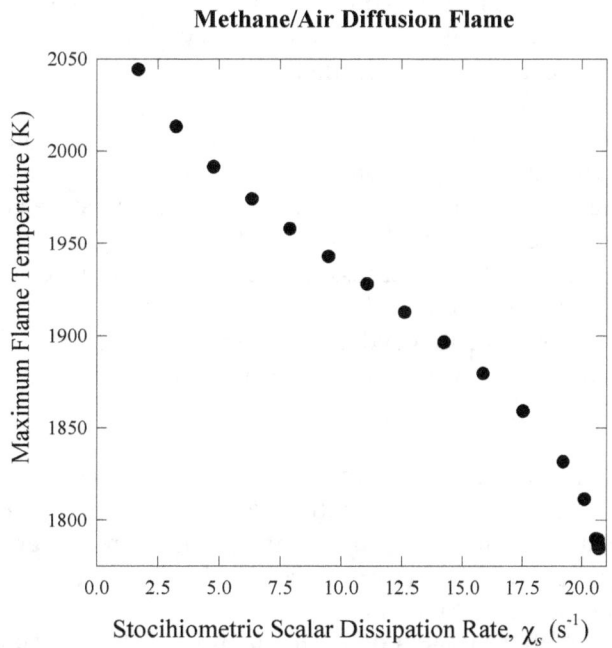

Figure 15. Values of maximum flame temperature for opposed-flow methane/air diffusion flames are plotted as a function of χ_s that is varied by changing the velocities of the fuel and air. Extinction is calculated to occur for $\chi_s = 20.7$ s^{-1}.

Table 10. Various Measures of Strain Rate Calculated for Methane/Air Diffusion Flames at Extinction

Source	L (cm)	u_f (cm/s)	u_o (cm/s)	a_g (s⁻¹)	$|a_o|$ (s⁻¹)	χ_s (s⁻¹)	T_{max} (K)
This Work (GRIMech 1.2)	1.0	157	-157	547	500	19.9	1792
This Work (GRIMech 1.2)	2.0	320	-320	558	509	20.7	1785
This Work (GRIMech 2.11)	2.0	312	-313	544	492	21.3	1793
This Work (GRIMech 3.0)	2.0	313	-313	546	500	21.7	1805
Dixon-Lewis et al. [88]	1.0	-	-	-	-	16.7	1769
Chelliah et al. [92]	1.0	185	-135	610	391	14.7	1758
Du and Axelbaum [117]	0.8	-	-	-	494	-	1770
Tanoff et al. [72]	-	-	-	-	350	-	1808
Tanoff et al. [72]	-	-	-	-	490	-	-
Tanoff et al. [72]					585	-	-
Chan et al. [101]	-	-	-	-	-	19.4	1773
Lentati and Chelliah [73]	1.0	-	-	-	425	-	1790

identified for methane/air diffusion flames. However, Yang and Kennedy have reported measurements of χ_s for a methane flame near inhibited by trichloroethylene near extinction. [116] These authors found that the extinction strain rate varied linearly with trichloroethylene volume fraction. Assuming that values of χ_s have the same dependence, it is possible to estimate a value of χ_s =18.6 s⁻¹ from their result for χ_s with 16.1 % trichloroethylene added to the methane fuel. Given the various approximations, this represents reasonable agreement with the current calculated value.

In order to obtain insights into the effect of the separation of the fuel and oxidizer sources, calculations were also performed for L = 1.0 cm. Table 10 compares the various measures used to quantify extinction for the two sets of calculations and includes the maximum flame temperature at extinction. While the initial flow velocities required for extinguishment depend strongly on L, the various measures for the effects of strain vary only slightly for the two cases. These minor differences are most likely due to the variations in the grids used for the calculations. These limited results suggest that for the assumed initial conditions, i.e., no radial gradients at the fuel and air exits, the velocities of fuel and air required for extinction are roughly proportional to L, while the various measures used to describe the role of strain are roughly independent of L.

It is of interest to compare the results of the current calculations with comparable modeling results found in the literature. Parameters used to characterize methane/air diffusion flames near extinction are included in Table 10 for a number of studies. Substantial variations are evident for calculated extinction parameters in terms of a_g, $|a_o|$, and χ_s. As discussed earlier, the GRIMech version used for the current study (Version 1.2) has been superseded by two later versions. A series of calculations was performed for methane/air flames using GRIMech 2.11 and 3.0 in order to assess whether substantial differences in calculated extinction behavior can be attributed to the GRIMech version used. Figure 16 compares the results for the three versions. Calculated values for u_o, $|a_o|$, and T_{max} for GRIMech Versions 1.2, 2.11, and 3.0 have been included in Table 10. The results for each mechanism fall close together, indicating that differences between the various GRIMech versions will have little effect on the findings discussed here.

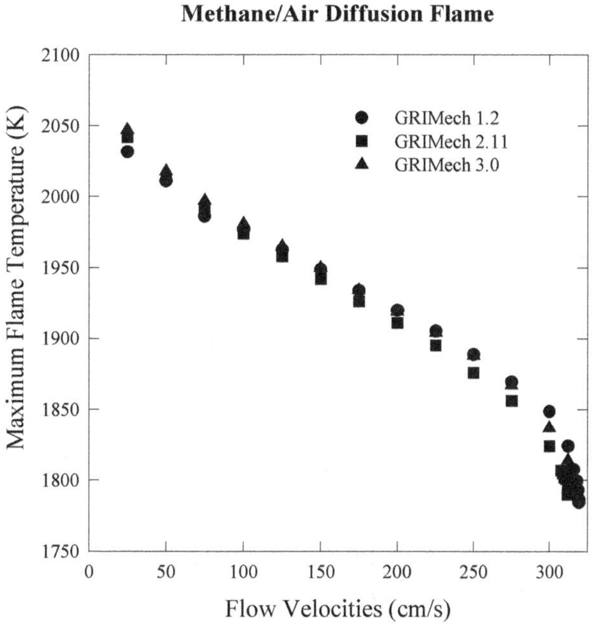

Methane/Air Diffusion Flame

Figure 16. Maximum flame temperatures are plotted as a function of the equal fuel and air velocity magnitudes for a methane/air opposed-flow diffusion flame. Results are compared for calculations using Versions 1.2, 2.11, and 3.0 of the GRIMech mechanism. [106]

The three sets of results for Tanoff et al. included in Table 10 provide a strong clue as to the reason for these variations. [72] The only change in their calculations was that different detailed chemical-kinetic mechanisms were used for each. Values of $|a_o|$ vary by nearly a factor of two. One of the mechanisms used by Tanoff et al. was GRI-Mech 2.11 ($|a_o|$ = 585 s^{-1}) that would be expected to provide similar results to the calculations reported here. The actual difference is roughly 15 %. We are not able to provide an explanation for the discrepancy, but do note that the calculations of Tanoff et al. included the effects of thermal diffusion and radiative heat losses that are not treated by the current calculations. The only other study which used GRI-Mech 2.11 was that of Chan et al. [101] Their value of χ_s = 19.4 s^{-1} is in reasonable agreement with the current result of 21.3 s^{-1}.

Bai et al. [103] have also reported flame structure calculations for methane/air diffusion flame using a reduced chemical-kinetic mechanism taken from Peters. [119] Unlike most earlier studies, they used the scalar dissipation rate at the peak temperature, χ_p, as their characteristic parameter. For a methane/air flame near extinction they calculated χ_p = 26.7 s^{-1}. The corresponding result from the current calculations is χ_p = 27 s^{-1}. While the agreement between these two modeling studies is very good, it must be kept in mind that the current calculations do overestimate the experimental value for a_o by roughly 20 %.

Based on the above discussion, we conclude that the current calculations for methane/air flames are consistent with existing literature results, but that there is a relatively large variation in the quantitative results due to the dependence of the results on the particular detailed chemical-kinetic mechanism used.

The extinction behavior for a propane/air flame has also been calculated using the mechanism of Babushok and Tsang. [70] Figure 17 shows a plot of maximum flame temperature versus the propane and air velocity magnitudes. Comparison with Figure 10 shows that while the flow velocities at extinction for the methane (320 cm/s) and propane (329 cm/s) flames are roughly the same, the maximum flame temperatures (1785 K and 1695 K, respectively) differ substantially. Differences of comparable magnitude exist over the full range of flow velocities considered.

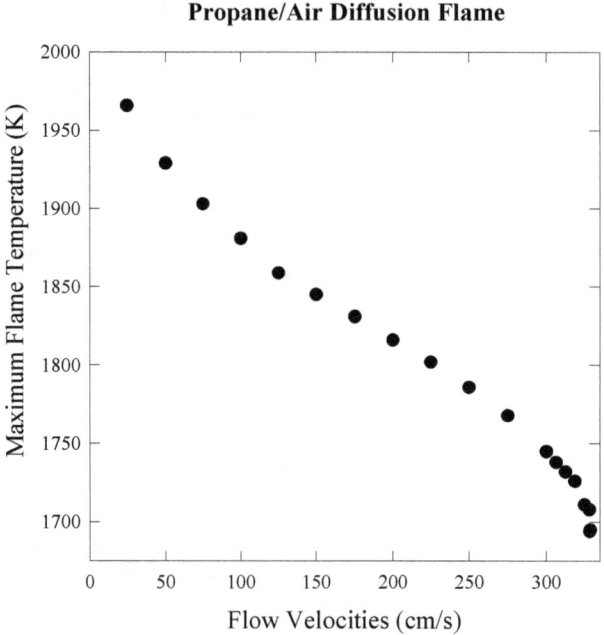

Figure 17. Maximum flame temperatures are plotted as a function of the equal fuel and air velocity magnitudes for a propane/air opposed-flow diffusion flame.

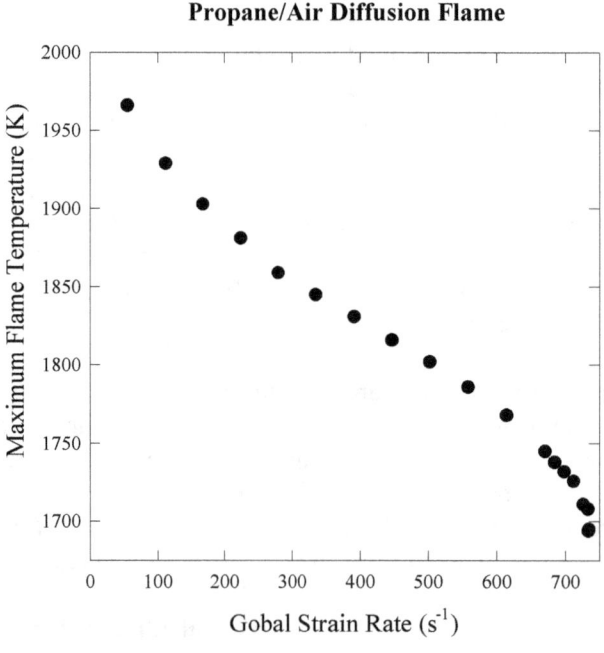

Figure 18. The maximum flame temperature is plotted as a function of the global stain rate a_g for a propane/air opposed-flow diffusion flame.

It is also of interest to compare the propane flame extinction behavior in terms of the global strain rate, local strain rate behavior on the oxidizer side, and the scalar dissipation rate at the stoichiometric point. Figure 18 shows a plot of the maximum flame temperature versus the global strain rate, a_g. Extinction is calculated to take place for $a_g = 735$ s^{-1}. Experimental values reported in the literature include those of Puri and Seshadri [114] ($a_g = 610$ s^{-1}), Lin and Faeth [120] ($a_g = 590$ s^{-1}), DeCroix and Roberts [98]

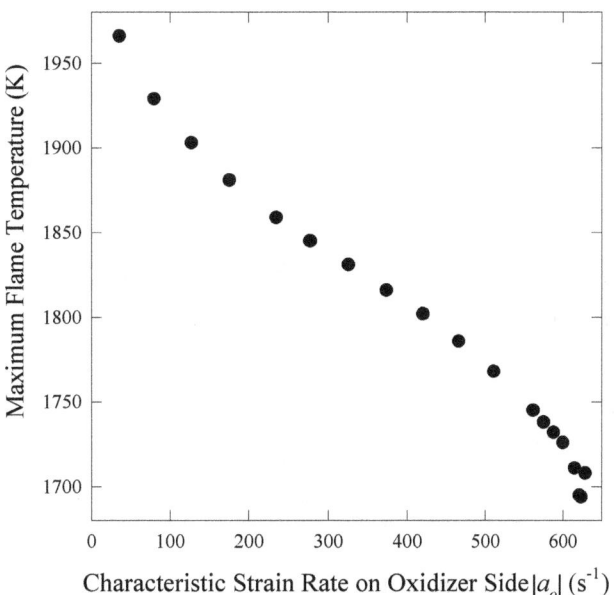

Figure 19. The maximum flame temperature is plotted against a characteristic local strain rate, $|a_o|$, defined to be equal to the highest absolute value of a on the air side observed outside of the boundary layer associated with the combustion zone for the propane/air flame.

(a_g = 700 s^{-1}) and Zegers et al. [52] (a_g = 381 s^{-1}). These values cover a wide range, with all of the experimental results falling below the calculated value. The reason for the wide variation in experimental results is not clear, but it may be associated with the different initial velocity profiles used in the various studies.

Figure 19 shows the calculated temperature data replotted in terms of the local strain rate $|a_o|$. Extinction is calculated to occur for $|a_o|$ = 621 s^{-1}. This result can be compared with experimental values available in the literature: Du et al. [115] ($|a_o|$ = 510 s^{-1} ± 10 s^{-1}), DeCroix and Roberts [98] ($|a_o|$ = 721 s^{-1}) and Zegers et al. [52] ($|a_o|$ = 583 s^{-1} ± 53 s^{-1}). As was found for a_g, there is a great deal of variation in the experimental results for $|a_o|$ that is greater than was found for methane flames. It is difficult to understand these variations since the local strain rate is supposed to compensate for differences in initial velocity profiles for the fuel and oxidizer. DeCroix and Roberts discuss experimental difficulties that arise when using propane for opposed-flow flame measurements. It is possible that these difficulties are responsible for the observed variations. Even though the experimental value is not well defined, it is encouraging that the calculated value falls in the general range provided by the experiments.

The calculated temperature data for a propane/air flame are replotted as a function of the mixture fraction scalar dissipation rate at the stoichiometric location, χ_s, in Figure 20. Flame extinction is calculated to occur when χ_s is increased to 26.3 s^{-1}. No other experimental or modeling values of this parameter have been identified for propane/air flames. However, Bai et al. have reported a calculated value of 66.0 s^{-1} for χ_p, the value of χ at the location of the peak flame temperature. [103] The value obtained from the current calculations is 44.9 s^{-1}, which must be considered poor agreement. Recall that the corresponding values were quite close for the methane/air flame.

The calculations and experiments suggest that propane burns more robustly than methane. Propane extinction requires a higher strain rate (calculated $|a_o|$ = 621 s^{-1} for propane versus $|a_o|$ = 509 s^{-1} for

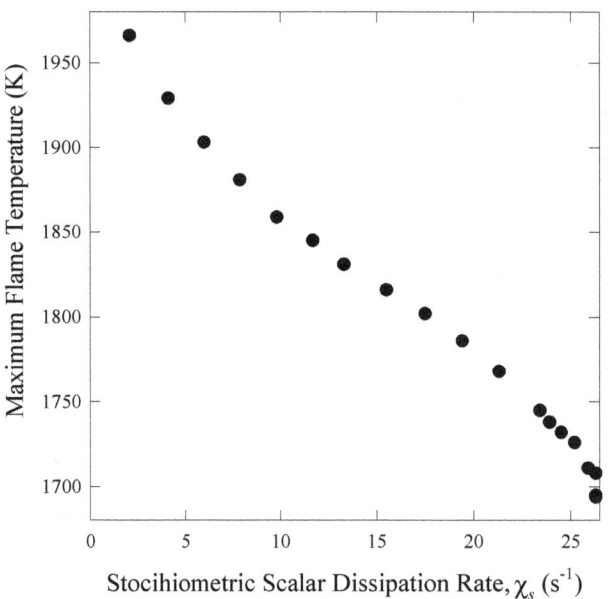

Figure 20. Values of maximum flame temperature for opposed-flow propane/air diffusion flames are plotted as a function of χ_s that is varied by changing the velocities of the fuel and air. Extinction is calculated to occur for $\chi_s = 26.3$ s^{-1}.

methane) and the calculated maximum flame temperature is lower for propane (1695 K for propane versus 1784 K for methane). With regard to strain rate, the difference is likely larger than indicated here since the value calculated for methane seems to be overestimated by roughly 30 %. The differences between methane and propane are consistent with the discussion of Maĉek. [42]

E. Extinction and Extinguishment of Methane and Propane Diffusion Flames by Nitrogen

The purpose of the detailed chemical-kinetic modeling calculations is to investigate the effects of thermal agents on the structure of opposed-flow diffusion flames and to use the findings to gain insights into flame extinguishment by these agents. Note that we are purposefully distinguishing between the meanings of flame "extinction" and "extinguishment". *Flame extinction* is used to describe a flame that goes out due to the local strain rate or other conditions causing the flame to become unstable. *Flame extinguishment* refers to the more specific conditions necessary to put out buoyancy-dominated flames that are expected to be characteristic of most fires. The cup burner test is an example of an experimental technique designed to characterize extinguishment. [46,47] Thus flame extinguishment can be considered to be a particular case of flame extinction.

In order to evaluate the effectiveness of the calculations, it is necessary to compare with experimental findings. Since the focus of this investigation is extinguishment of fires, the most appropriate measure to consider for characterizing diffusion flame extinction is the amount of thermal agent required to extinguish a buoyancy-dominated flame burning at standard gravity. This concentration is related to the concept of a limiting oxygen index (LOI), which is defined to be the lowest oxygen concentration in air diluted with nitrogen that can support a flame burning under these conditions. The minimum agent volume fraction required to meet this condition is sometimes referred to as the inerting concentration.

In order to define an extinguishing condition for the addition of thermal agents to the oxidizer it is necessary to relate the calculations to an experimental result. For the methane/air and propane flames of interest here we wished to use the LOI for this purpose. Somewhat surprisingly, only a limited number of LOI values have been published for these flames.

For methane, Simmons and Wolfhard reported a LOI value of 0.139 using a hemispherical burner located in a slow oxidizer counter flow. [41] This corresponds to an added N_2 volume fraction of 33.8 %. Measurements by Ishizuka and Tsuji in a similar burner yielded a value of 31.9 %. [48] Puri and Seshadri reported that a 28.6 % N_2 volume fraction was sufficient to extinguish an opposed-jet methane/air diffusion flame. [114] In a recent, as yet unpublished, study, researchers at NIST made measurements in a similar low strain-rate flame and measured a value of 30.0 % for the maximum N_2 volume fraction at extinguishment. [121] At the 1999 HOTWC meeting, Ural reported a study of extinguishment of a methane flame in a cup burner apparatus. [57] The extinguishing volume fraction of N_2 over a relatively wide range of fuel and oxidizer velocities was 27.1 %.

As for methane, there are only a limited number of LOI values reported for propane/air diffusion flames. Simmons and Wolfhard measured a LOI of 0.127 using their porous hemispherical burner, corresponding to a nitrogen volume fraction for extinguishment of 39.5 %. [41] Note that that the measured extinguishing concentration is roughly 17 % greater than observed for a methane flame. For an opposed-flow diffusion flame, Puri and Seshadri recorded an extinguishing volume fraction of 34.8 % nitrogen. [114] This value is nearly 22 % higher than found for methane by the same authors. Note that for a given fuel, the nitrogen extinguishing volume fraction is significantly lower for the opposed-flow burner.

Several values of nitrogen extinguishing concentration are available for propane flames burning in a cup burner configuration. Hamins et al. reported an extinguishing volume fraction of 32.5 %. [33] Babb et al. measured a value of 33 %. [56] The results of Ural also yield a value of roughly 33 %. [57] These results are all very close together, suggesting that lab-to-lab variations are not responsible for the differences observed between different burners. Comparison of the results of Babb et al. for methane and propane indicates that a propane flame again requires a higher concentration for extinguishment (roughly 22 %).

The range of extinguishing concentrations for methane and propane flames burning in air for different types of burners reported in the various investigations cited above is much larger than can be explained by uncertainty in the measurements and suggests that different types of diffusion flames yield different results. This raises the important question of what type of burner is appropriate for determining a characteristic extinguishing nitrogen volume fraction (or LOI) for buoyancy-dominated diffusion flames.

Due to the limited data available and the wide range of values reported, it was decided to carry out additional measurements of nitrogen extinguishing concentration using two diffusion flame burners available at NIST. The first was a modified version of the DLAFSS system described earlier that incorporates a Tsuji-type opposed-flow cylindrical diffusion flame burner, and the second was a Santoro-type burner (a coflow diffusion flame) previously used at NIST for investigations of diffusion flame structure. Descriptions of these burners are provided in the following section.

1. Experimental Systems for Measuring Extinguishing Concentrations of Gaseous Thermal Agents

    a) Modified Dispersed Liquid Agent Fire Suppression Screen System (Tsuji Burner)

The DLAFSS described in Section (II.B.1.b) was used to investigate the extinguishment of Tsuji-type methane and propane diffusion flames by gaseous thermal agents. Even though designed primarily for evaluating liquid fire suppression agents, the system can also be used for gaseous agents. [35,36] The injection point for gases is at the bottom of the system as can be seen in Figure 6.

The goal of these measurements was to identify extinction conditions for thermal agents added to the oxidizer flow for porous-burner-stabilized opposed-flow diffusion flames under the relatively low strain rates characteristic of buoyancy-dominated flames. Typically, the Tsuji burner is operated at considerably higher oxidizer flow velocities than required here, and the resulting strain rates are much greater than found in buoyancy-dominated flames. In order to generate the required low oxidizer velocities, the fan used to induce the flow in the DLAFSS was operated at the lowest setting possible. The thermal agents were then added to the induced air flow. Note that the oxidizer flow velocity is expected to vary as the agent volume fraction increases.

The flow rate of agent was controlled using a needle valve, and the volume flow rate was measured using a rotameter (Fischer Porter, Model 10A1755). Even though the agent volume flow rate was known, it was discovered that it was not possible to predict the agent volume fraction using only the assumed air-flow rate. This observation was attributed to variations in the oxidizer total volume flow rate due to changes in the fan efficiency as agent was added. An alternate approach was adopted to estimate the oxidizer total flow rate and thus the local flow velocity at the flame. The agent concentration was determined by measuring the oxygen volume fraction in the oxidizer stream at the center of the settling chamber of the wind tunnel (see Figure 6) using a paramagnetic oxygen meter (Servomex Oxygen Analyzer Model 570A) in conjunction with a small sampling pump (Gilian Hi Flow Sampler, Model HFS 513A). The oxygen analyzer was zeroed and spanned using nitrogen and room air. Since the volume flow rate for the agent was known, knowledge of the degree of air dilution allowed the total flow rate to be calculated assuming complete mixing.

In a typical DLAFSS experiment, extinction is defined to occur when the flame on the upstream side of the burner blows out. For high strain-rate conditions, it is generally found that a flame stabilized in the wake of the cylinder continues to burn following blow out of the upstream flame. During the current experiments, it was observed that when sufficient agent was added to the air flow to induce extinction of the stretched flame on the upstream side of the cylinder, the downstream flame was either extinguished at the same time, or disappeared when the agent concentration was increased only slightly. Since the stabilized flame is essentially a buoyancy-dominated flame, this is taken as evidence that the measurements are actually recording flame extinguishment values.

Extinguishment tests were performed by gradually adding the gaseous inhibitor to the oxidizer (air) stream until blow-off occurred (abrupt transition from envelope to wake flame or blow out). The volumetric flow rates of the suppressant at blow-off were recorded. The oxygen concentration was then noted for the same flow conditions.

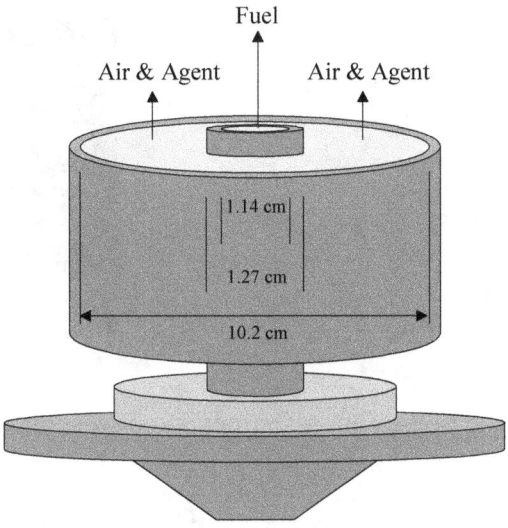

Figure 21. A schematic of the modified Santoro burner is shown.

b) NIST Coflow Diffusion Flame Burner (Santoro Burner)

An existing coflow burner and computer-controlled gas mixing system were modified to allow extinguishing concentrations of thermal agents to be determined for buoyancy-dominated laminar diffusion flames formed by a flow of gaseous fuel from a circular tube into a slow surrounding coflow of an air/agent mixture. The burner is based on a design originally developed by Santoro et al. that is commonly referred to as a "Santoro burner". [122] This type of burner has been employed for numerous flame studies at NIST. In a particularly relevant study, Smyth and Everest reported volume fractions of CF_3Br and CF_3I in air required to extinguish a propane flame on the Santoro burner. [123]

The actual burner used in this investigation is a modified Santoro burner that was originally employed to investigate acoustically forced, time-varying diffusion flames. [124] A drawing of the burner is shown in Figure 21. It consists of an open 13.7 cm length of tubing with a 1.27 cm outer diameter that serves as the central fuel tube. The tube lip thickness is 0.065 cm, yielding an inner diameter of 1.14 cm. One end of the tube is attached to a plenum containing a loud speaker. The speaker, used to acoustically force the fuel flow, was not employed during this study. The open end of the fuel tube extends 0.4 cm above a surrounding 10.2 cm diameter ceramic honeycomb formed from 0.15 cm square cells and having an overall length of 2.5 cm. After passing through a chamber filled with glass beads and several screens, a flow of air and agent enters the honeycomb to form the surrounding coflow for the flame.

The flow velocities for the fuel and coflow streams were low and resulted in small buoyancy-dominated flames that were highly susceptible to room drafts. Screens were placed around the burner to minimize these disturbances. Flame products were removed from the area by an exhaust cone connected to the room exhaust by a flexible hose. The effect of the exhaust cone on the flames was judged to be negligible due to its low rate of suction and its location well above the burner.

Fuel and the air/agent mixture were delivered to the burner using a system of thermal mass flow controllers (MFC's). The flow-control system, shown in Figure 22, is a modification of a system developed at NIST for a previous investigation as described in detail by Pitts et al. [125] The flow system was software controlled by a desktop 486-DX personal computer. The volume flow rates required to

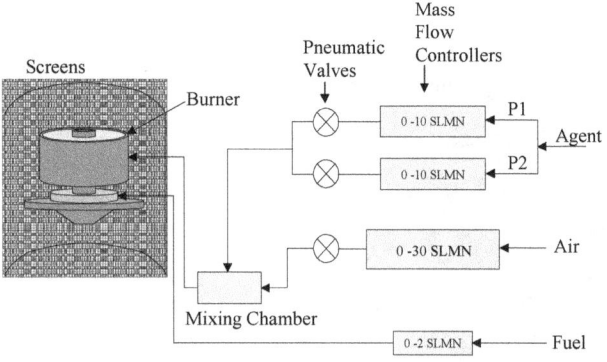

Figure 22. A schematic and photo of the Santoro burner and flow control system are shown.

generate precise mixtures of air and agent were specified by the software. The air and agent streams were combined upstream of the burner in a mixing chamber designed to ensure a homogenous mixture. Pneumatic shut-off valves prevented unwanted backflow of gases and allowed for easy calibration of each MFC. Modifications for the current study included the addition of a 30 SLMN (standard liters of nitrogen per minute) mass-flow controller for the air flow and moving the 10 SLMN and 2 SLMN mass-flow controllers to the agent and fuel streams, respectively. Appendix B includes additional details concerning the MFCs and their calibration. Gas supplies for the fuel and agents were bottled gases with nominal purities of 99.0 %. Air was supplied from an in-house high-pressure line, and its relative humidity was a consistent 4.5 %.

The system performed automated experimental control and data acquisition using a Flash-12 Model 1 data acquisition card and an ACAO-12-2 analog and digital output board, both from Strawberry Tree, to provide interfaces with the experiment. The mass flow controllers were connected to a MKS Instruments, Inc. Type 247C 4-channel readout, set point, and power supply that also served to interface the MFC's with the data acquisition and output boards.

Software control was provided by an icon-based worksheet, developed using WorkBench PC (from Strawberry Tree, Inc.). The worksheet, named AGFCTL.WBB (Automatic Gas Flow Control), was a modification of a similar program used in the study described by Pitts et al. [125] The modified worksheet provides the necessary calculations and controls for operation of the mass-flow controllers, on/off control of the electro-pnuematic valves and indicator lights, series step changes in agent concentrations, and display and data logging of key parameters. Another worksheet, MFCCAL.WBB (Mass Flow Controller Calibration), was developed for the calibration of the individual mass flow controllers.

Most experimental systems designed to study the effects of a fire-fighting agent on coflow diffusion flames, e.g., the cup burner, create agent/air mixtures by holding the air volume flow rate constant and adding the agent. As a result, the velocity of the oxidizer flow increases as the agent is added. This increases the local strain rates in the vicinity of the flame and is therefore a potential source of variation in the measurement. An advantage of the computer-controlled system used here is that it is possible to maintain the total volume flow rate for the agent/air mixture constant as the agent concentration increases. For a majority of the methane flame measurements reported here, the nominal exit flow velocities for the fuel and oxidizer coflow were matched at 7 cm/s. Similar velocities were employed in earlier studies with this burner. The required volume flow rates of fuel and oxidizer were calculated using the general equation,

$$Q_{exit} = U_{exit} A_{exit}, \qquad (15)$$

where U_{exit} is the desired velocity and A_{exit} is the flow area and equals 1.02 cm² and 79.8 cm² for the fuel and air, respectively. Other combinations of fuel and coflow velocities were employed during the study as discussed below.

An experiment consisted of:
1) Entering the desired volume flow rates for fuel and oxidizer in SLM into the appropriate worksheet locations.
2) Setting the initial volume fraction for the desired thermal agent at the computer control panel window. The starting volume fraction was typically 0.010 to 0.020 less than the expected extinguishment concentration. (Preliminary survey experiments were run to determine approximate extinguishing concentrations)
3) Starting the program at the computer-control panel. The program automatically increased the agent concentration in the agent/air mixture by increments of 0.001 in volume fraction. The agent concentration was held for 40 s in order allow any effects of transients associated with the volume fraction change to decay.
4) Monitoring the flame and recording the agent concentration at which the flame extinguished. Once the flame was extinguished, the program was immediately manually aborted, which in turn interrupted the gas flows.

Additional details concerning the calculation of the appropriate gas volume flow rates are included in Appendix B. A discussion of the analysis employed to estimate uncertainties for the volume fraction of agent in air is also included in Appendix B.

2. Results for Thermal Agent Extinction of Methane and Propane Diffusion Flames Burning on Tsuji and Santoro Burners

The extinguishment behavior of flames stabilized on the Tsuji and Santoro burners were investigated using the gaseous thermal agents nitrogen, carbon dioxide, argon, and helium. The following two sections summarize the general observations and the results of extinguishment volume fraction measurements.

a) Tsuji Burner

Multiple extinguishment tests were run for each agent. Table 11 summarizes measured values of extinguishing volume fraction for the four thermal agents. The repeatability of the measurements is indicated by the magnitude of the standard deviation. The actual number of repeated tests is denoted by the symbol n. The table also includes values of the stagnation velocity gradient, a_T, defined as $2 V_o/R$, where V_o is the velocity based on the total volumetric flow rate of air and gaseous agent at blow-off and R is the burner radius. a_T represents the characteristic stagnation strain rate that is often used for Tsuji burners and has units of s⁻¹. a_T is similar to the global strain rate, a_g (see Eq. (14)), used to characterize the strain rate for axisymmetric opposed-flow diffusion flames. Note that a one-to-one correspondence should not be expected for a_g and a_T, and direct comparisons are inappropriate.

As already discussed, measurements of extinguishing volume fractions in the literature are limited for opposed-flow porous burners. Simmons and Wolfhard [41] reported values of 34 % and 39.5 % for the extinguishing concentrations of methane and propane flames on a hemispherical porous burner, while

Table 11. Extinguishment Results Using the DLAFSS

Agent	Methane		Propane	
	Agent volume fraction at blow-off	$a_T = 2V_o/R$ (s^{-1})	Agent volume fraction at blow-off	$a_T = 2V_o/R$ (s^{-1})
Carbon Dioxide	0.240 ± 0.003* (n = 4)	39 ± 1	0.257 ± 0.008 (n = 3)	45 ± 2
Nitrogen	0.332 ± 0.004 (n = 5)	52	0.390 ± 0.004 (n = 5)	56
Helium	0.207 ± 0.003 (n = 5)	94	0.224 ± 0.003 (n = 5)	95 ± 1
Argon	0.468 ± 0.002 (n = 4)	44 ± 1	0.550 ± 0.005 (n = 5)	44
* mean ± standard deviation n = number of samples				

Ishizuka and Tsuji [48] found a value of 31.9 % for a methane flame stabilized on a cylindrical porous burner. These values agree very well with those included in Table 11, (33.2 ± 0.4) % and (39.0 ± 0.4) % for methane and propane, respectively. This agreement is found even though two different types of porous burners were used. Recall that the extinguishing nitrogen volume fractions of Simmons and Wolfhard were the highest observed for the various types of diffusion flames discussed earlier. It must be concluded that a flame for a given fuel is more stable on an opposed-flow porous burner than for the corresponding flames burning between two axisymmetric opposed-jet flows or in a coflow configuration. The reasons for these observations are currently unclear.

Note that in the case of helium included in Table 11, the values for $a_T = 2V_o/R$ are almost a factor of two higher than those obtained using nitrogen, carbon dioxide, and argon. Recall that the fan in the DLAFSS is set to operate at the lowest possible revolution rate. Even though the fan speed is nominally constant, when the density downstream of the fan is significantly reduced, the fan generates a higher volume flow rate. The net result is that the characteristic strain rate a_T, at the flame is roughly a factor of two higher for the helium case than for the other agents, and it is to be expected that the observed extinguishing concentration will be somewhat lower than would be measured for a purely buoyancy-dominated flame. The variation of extinguishing volume fractions with different agents will be discussed in more detail below.

b) Santoro Burner

The coflow gas burner is a deceptively simple flame configuration. During the extinguishment studies summarized here a number of complex behaviors were observed that complicated what should have been, in principle, a simple measurement of extinguishing concentration. The behaviors are also shown to be fuel dependent, being different for methane and propane flames.

Consider first the methane base flame (i.e., fuel and oxidizer nominal velocities equal to 7 cm/s) burning in air diluted with N_2. With only air present, a blue flame was evident slightly below the exit and very close to the fuel tube. Further downstream the flame was yellow indicating the presence of soot. As the N_2 concentration in the oxidizer was increased, the base of the flame gradually moved downstream and away from the burner tube, but the flame still appeared to be attached. At the same time, the fraction of the flame that appeared yellow decreased. When the N_2 volume fraction reached 20.2 %, an up and down oscillation of the flame base suddenly appeared as indicated in Figure 23. This flame was entirely blue. The oscillating flame only existed over a narrow range of N_2 volume fraction. Similar oscillations for flames near extinction have been reported previously in the literature. [126,127]

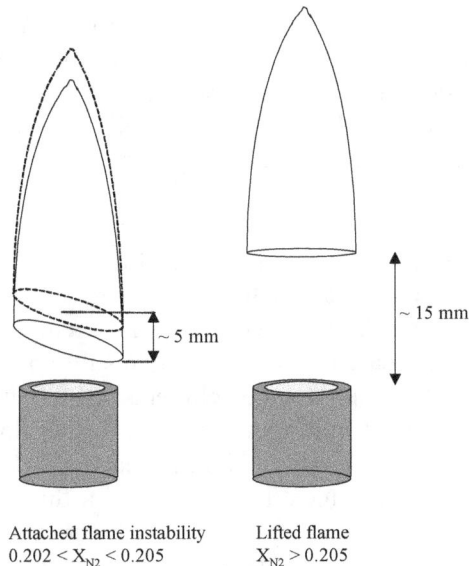

Figure 23. Sketch of a methane flame showing the attached flame oscillation and subsequent stabilized lifted flame that result from the addition of nitrogen to the air coflow.

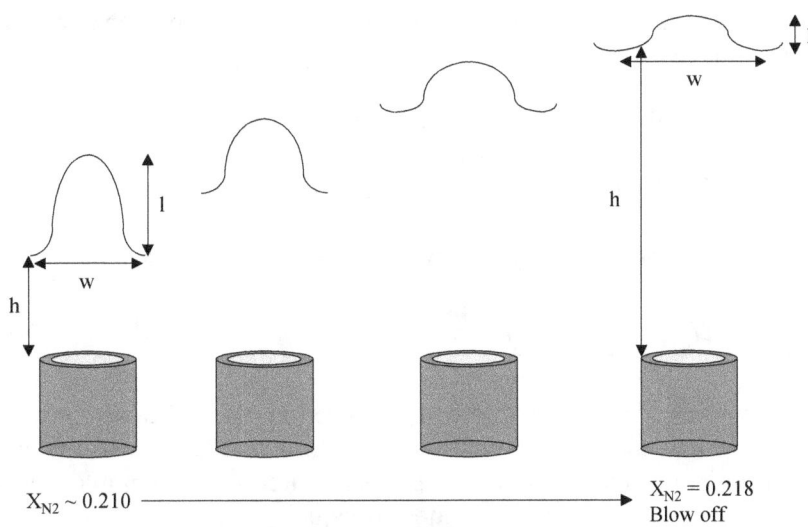

Figure 24. The evolution of the methane diffusion flame structure as additional N_2 is added to the coflow air stream is shown.

When the N_2 volume fraction was increased to 20.5 %, the flame detached from the burner and moved downstream, where it became stable and floated lazily up and down 15 mm to 20 mm above the burner. With further increases in N_2 concentration, the flame moved further downstream, while becoming shorter and broader. These changes in appearance are shown in cartoon form in Figure 24. Similar lifted-flame structures have been observed in investigations with different fuels and burner configurations and are attributed to the presence of triple flames. [128,129,130,131] The presence of lifted flames has also been reported during cup burner experiments using gaseous fuels. [33,56,57] When the N_2 volume fraction reached 21.8 %, the lifted flame could no longer be stabilized, and it simply floated away, i.e., it was blown off.

Increasing N_2 dilution of the air resulted in three distinct behaviors—attached flame instability, lift off, and blow off. The N_2 volume fraction range where these processes occurred was narrow, ranging from 20.5 % to 21.8 %. It is difficult to define any of these processes as a true extinguishment, though blow off would seem to be the most appropriate. For reasons that will be clear shortly, the onset of the fluctuation behavior was used for the characteristic extinguishing concentration for methane flames burning in air diluted with thermal agents.

When CO_2, Ar, and He were used as diluents, the response of the baseline methane flame changed from that observed for N_2. Initial changes in flame structure with increasing inert agent volume fraction were similar to those observed with nitrogen. For all three agents, an agent volume fraction was reached for which the attached flame instability appeared. Similar to the results for N_2, additional small increases in the volume fractions of CO_2 and Ar resulted in detachment of the flame from the burner, with stabilization further downstream. At first, these flames moved further downstream with additional increases in diluent concentration, but eventually the flames reached lift-off heights where further increases appeared to have little effect. The most likely reason for this observation is that the flames reached a downstream location where surrounding room air was able to diffuse through the oxidizer coflow and provide sufficient O_2 to sustain combustion. For these cases it is impossible to define the agent volume fraction required for flame blow off. The behavior of the flame burning in air diluted with He was quite different. After a few oscillations following the onset of the attached flame instability, the flame was extinguished without further increases in helium concentration. Apparently, the onset of the attached flame instability was sufficient to extinguish the methane flame burning in the He/air mixture. Since the appearance of the attached flame instability was the only common characteristic for all four of the agents, its onset was selected as the condition determining extinguishment of methane flames.

When the baseline fuel and oxidizer volume flow rates for the methane case were used for propane, the resulting flame length was much greater than observed for methane. This observation is easily understood by considering the flame stoichiometries for the two fuels,

Methane: $\quad CH_4 + 2O_2 \rightarrow 2H_2O + CO_2$

Propane: $\quad C_3H_8 + 5O_2 \rightarrow 4H_2O + 3CO_2$,

which indicate that combustion of a given volume of propane requires 2.5 times more oxidizer than the same volume of methane. Since the entrainment rate of oxidizer into laminar diffusion flames is roughly independent of the fuel flow rate, a longer flame length is required for the complete combustion of propane than for methane when the volume flow rates are equal.

In order to have roughly comparable flame lengths for the two fuels, the nominal fuel velocity for propane was reduced to 40 % of that used for methane (i.e., a nominal velocity of 2.8 cm/s, volume flow rate of 0.171 L/min). With the oxidizer nominal velocity maintained at 7 cm/s, the propane flame length was comparable to that for the baseline methane flame. These flow velocities were adopted as the baseline conditions for the propane flame.

Similar to the baseline methane flame, the propane flame was attached to the fuel tube prior to the addition of nitrogen to the coflow. As the nitrogen volume flow rate was increased, the blue region at the base of the flame occupied a larger volume, as was also observed for the methane flame. However, unlike the methane flame, the base of the propane flame broadened substantially as it moved away from the burner tube with increasing nitrogen addition. By the time the nitrogen volume fraction was increased to 26.5 %, the flame base was stabilized approximately 5 mm downstream of the fuel exit. When the nitrogen volume fraction was increased to 27.4 %, the flame base was approximately 10 mm above the

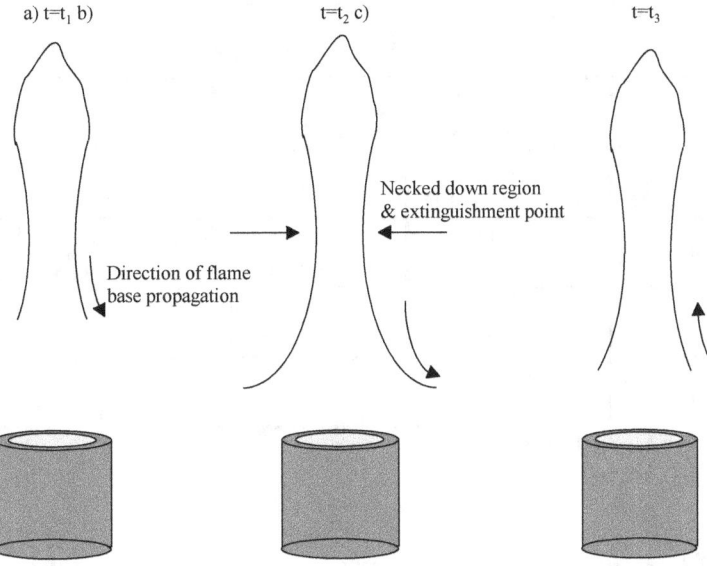

Figure 25. A sketch of the time evolution of a nearly extinguished propane flame is shown for the case of nitrogen dilution of the coflow air, $t_1 < t_3$.

fuel tube and began to oscillate with a combined vertical and radial motion. Further downstream the flame formed a necked-down region. With only an additional increase of 0.1 % nitrogen volume fraction, the flame suddenly extinguished by collapsing at the necked-down region. A cartoon of the nearly extinguished flame is shown in Figure 25. The extinguishment of the propane flame following stands in sharp contrast to the methane flame, which formed a stable lifted flame following detachment from the burner.

Unlike the baseline methane flame, the general response of the propane flame to dilution of the coflow air by CO_2, Ar, and He was the same as described above for nitrogen dilution. Since an easily characterized extinguishment was observed for each agent using propane, this event was designated as the characteristic flame extinguishment behavior.

The results of extinguishment studies for baseline methane and propane diffusion flames on a Santoro burner are summarized in Table 12. Agent volume fractions corresponding to the onset of flame fluctuation and flame extinguishment (if observed) are included. Expanded uncertainties, U, calculated as described in Appendix B are included in the table. Extinguishment was not observed for the methane flame with CO_2 and Ar. The volume fraction of agent required to induce oscillation is taken as the characteristic extinguishing concentration for the methane flame, while the actual concentration at flame extinguishment is used for the propane flame. Note that the small differences observed between the two values for given agent and fuel suggests that small differences will be introduced by the use of different definitions for extinguishment.

The extinguishing concentrations for a given agent listed in Table 12 for the Santoro burner are much lower than for the corresponding agent included in Table 11 for the Tsuji burner. A potential explanation for this observation is a dependence of the Santoro burner results on the fuel and/or flow velocities. In order to check this possibility, measurements were made using both methane and propane in which the velocities were varied from the baseline cases. These results are summarized in Table 13 and Table 14, which list results for flames where the fuel and oxidizer volume flow rates were varied independently. Recall that baseline values for the methane and propane flames are Q_{fuel} = 0.429 L/min and Q_{coflow} =

Table 12. Measured Agent Volume Fractions at Onset of Flame-Base Instability (Osc) and for Flame Extinguishment (Ext) of Baseline Methane and Propane Coflow Diffusion Flames

	Methane		Propane	
	Osc	Ext	Osc	Ext
Carbon Dioxide	0.120 ($U = \pm 0.001$)	-	0.169 ($U = \pm 0.009$)	0.176 ($U = \pm 0.004$)
Nitrogen	0.202 ($U = \pm 0.006$)	0.218 ($U = \pm 0.006$)	0.276 ($U = \pm 0.010$)	0.280 ($U = \pm 0.016$)
Helium	0.200 ($U = \pm 0.004$)	0.200 ($U = \pm 0.004$)	0.271 ($U = \pm 0.007$)	0.273 ($U = \pm 0.006$)
Argon	0.290 ($U = \pm 0.004$)	-	0.356 ($U = \pm 0.006$)	0.368 ($U = \pm 0.007$)

U = Expanded Uncertainty with K = 2

Table 13. Effects of Fuel and Oxidizer Velocity on Agent Concentrations at Onset of Flame Base Oscillation (Osc) and Flame Extinguishment (Ext) for the Methane Coflow Flame

Q_{fuel} L/min	Q_{coflow} L/min	CO_2 Osc	Ext	N_2 Osc	Ext	He Osc	Ext	Ar Osc	Ext
0.429	33.52	0.120	-	0.202	0.218	0.200	0.200	0.290	-
0.215	33.52	0.121	0.128		0.198	0.198	0.291	-	
0.107	33.52	0.123	0.131	0.204	0.233	0.199	0.199	0.296	0.309
0.429	16.76	0.124	-	0.205			0.206	0.294	-
0.429	8.38	0.127	-	0.206		-	0.232	-	0.301	-

Table 14. Effects of Fuel and Oxidizer Velocity on Agent Concentrations at Onset of Flame Base Oscillation (Osc) and Flame Extinguishment (Ext) for the Propane Coflow Flame

Q_{fuel} L/min	Q_{coflow} L/min	CO_2 Osc	Ext	N_2 Osc	Ext	He Osc	Ext	Ar Osc	Ext
0.171	33.52	0.169	0.176	0.276	0.280	0.271	0.273	0.356	0.368
0.086	33.52	0.171	0.174	0.270	0.274	0.280	0.280	0.365	0.370
0.086	16.76	0.172	0.176			0.296	0.297	0.362	0.363
0.429	33.52	0.149	0.178	0.228	0.280				

33.5 L/min and Q_{fuel} = 0.171 L/min and Q_{coflow} = 33.5 L/min, respectively. Blank spots indicate measurements that were not recorded for a particular agent, while dashes indicate cases where blow off of the methane flame was not observed.

In general, the volume fractions of added agent where flame oscillation and extinction occur are nearly insensitive to the fuel and oxidizer velocities for the ranges considered. The only exceptions are the flames for which helium was added to the air. For the methane flame, the effect of the velocities on the helium results is only evident for changes in the oxidizer velocity, with the extinguishing volume fraction increasing by 15 % as the velocity is decreased by a factor of four. In the case of propane, the dependence of the helium extinguishing volume fraction on flow velocity is evident for both the fuel and

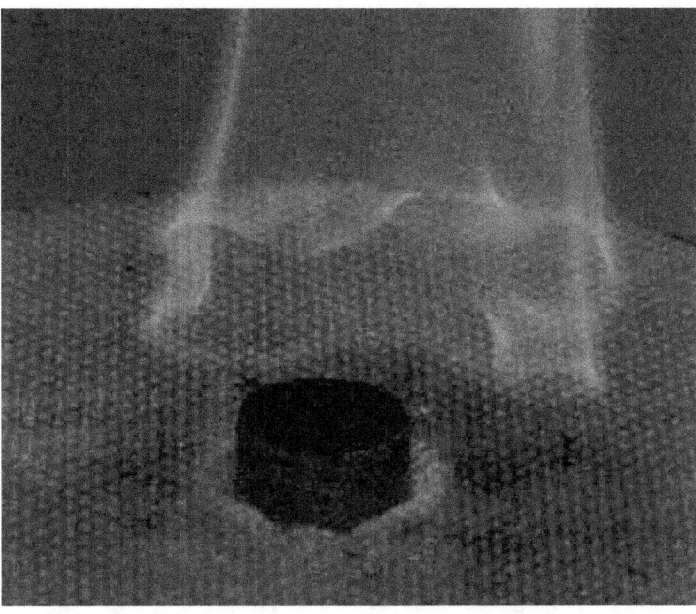

Figure 26. A time-resolved photograph taken by a digital camera shows the oscillating propane flame burning in air diluted with nitrogen. The complex azimuthal structure of the flame base, which has a much larger area than the burner tube, is clear.

oxidizer. A reduction of the oxidizer velocity by a factor two resulted in a 10 % increase in the volume fraction of helium required to induce flame oscillation or extinguishment. Based on the relative insensitivity to the fuel and oxidizer velocities, we conclude that the experimental results provide good estimates for the agent extinguishing conditions using the Santoro burner.

Generally, the variations in fuel and oxidizer velocities from the baseline cases are reductions. This is appropriate because the net effect should be a reduction in strain rate at the flame surfaces due to the initial flow velocities. The low insensitivity of the agent extinguishing volume fractions to these changes indicates that these strain rates are small relative to those resulting from the buoyancy-induced flows associated with the heated combustion gases.

The only case where the nominal fuel velocity was increased from the baseline value was for the propane flame, where it was set to the value for the methane baseline flame, i.e., a volume flow rate of 0.429 L/min. Recall that the resulting flame length is much longer than for the corresponding methane flame. The primary effect of this increase was a dramatic decrease in the volume fractions of CO_2 and N_2 required to induce flame oscillations. The extinguishing concentrations were nearly unchanged.

Not only did the propane flame oscillations begin at much lower concentrations for the higher fuel-flow velocity, but the flames present during the oscillation also displayed unique instability behaviors over the agent concentration range where the oscillation occurred. Similar to the smaller propane flames, a combined vertical and radial oscillation with a necked down region above the flame base appeared at the first occurrence of the oscillation. However, as the agent concentration was increased toward the extinguishment value, the vertical and radial displacements grew in intensity, while the frequency of oscillation decreased. The flame base became distorted, losing its axisymmetric symmetry, and developed lobes or flower-like petals, which increased in number with increasing agent concentration. Figure 26 shows a digital photograph recorded for a propane flame as it approached extinction due to the addition of nitrogen to the air coflow. Near flame extinction, as many as six lobes or petals were

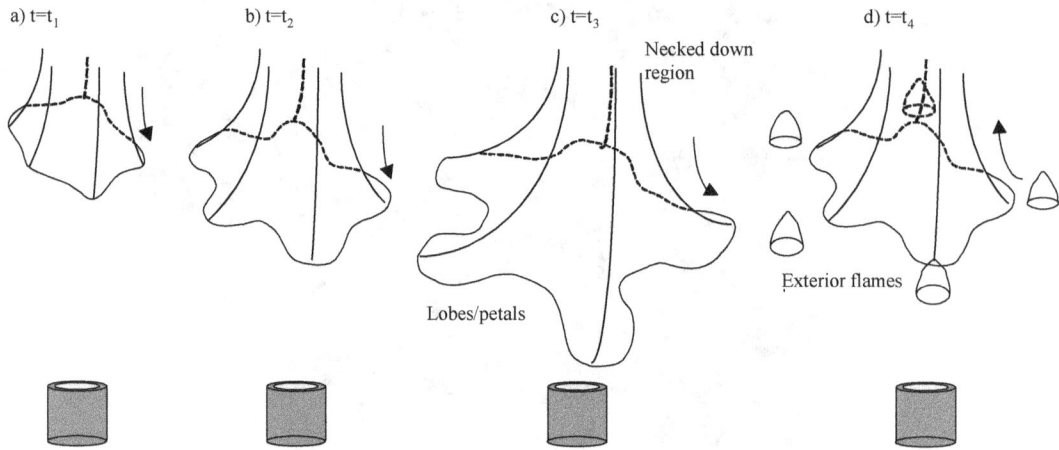

Figure 27. The cartoon shows the time evolution of the pulsing instability in the large flow-velocity propane diffusion flame observed when a concentration of a thermal fire-extinguishing agent near that required for extinguishment is added to the air coflow. The arrows indicate the direction of flame base propagation. The view is from an angle above the flame.

observed. At times the ends appeared to detach from the main flame and form small isolated flame tips. A cartoon of the various flame instability behaviors observed with increasing agent concentration is shown in Figure 27.

The physical mechanisms responsible for these interesting flame instability behaviors are not currently known and are not legitimately part of the current investigation. However, it is worthwhile to point out the similarities between these flame structures and the structures formed by vortices in the near fields of axisymmetric jets and flames subject to strong forcing or absolute instability. [132,133,134]

The relative insensitivity of the measured characteristic extinction concentrations for thermal agents to the flow velocities used in the Santoro burner indicates that they are indeed extinguishment volume fractions. The results have the same dependence on fuel as already discussed, with characteristic agent volume fractions required for extinguishment varying from 25 % to 46 % greater for the propane flames.

Table 15 provides a summary of nitrogen extinguishing volume fractions reported in the literature for various methane and propane diffusion flames, along with results from the current investigation. Values for different types of diffusion flame burners are grouped together. The dependencies of extinguishing concentration on both fuel and burner type are evident.

It is clear from the results in Table 15 that N_2 extinguishing volume fractions for the Santoro burner are considerably lower than have been reported for cup burner studies. This is, at first, somewhat surprising since both burners generate coflow diffusion flames, and the flow velocities of fuel and oxidizer are comparable in each, with the results for both types of flame shown experimentally to be nearly independent of the initial flow velocities. There are three major differences between the cup and Santoro burners. First, cup burners typically have a glass "cup" which serves as the fuel port, while the Santoro burner uses a straight stainless steel tube. Second, the inside diameters of the fuel tube (2.2 cm to 2.8 cm) for the cup burners included in Table 15 are on the order of a factor of two greater than for the fuel tube (1.14 cm) in the Santoro burner. Third, the coflow in the cup burner is enclosed within a tube (open at the top) while the Santoro burner coflow is exposed to the ambient surroundings.

Table 15. Extinguishing Nitrogen Volume Fractions For Various Diffusion Flames Burning Methane and Propane

Burner Type	Methane	Propane
(Porous Hemisphere/Opposed Flow) [41]	33.8 %	39.5 %
Tsuji (Porous Cylinder/Opposed Flow) [48]	31.9 %	--
Tsuji (Porous Cylinder/Opposed Flow) (This Work)	32.2 %	39.0 %
Opposed-Flow Gas Jets [114]	28.6 %	34.8 %
Opposed-Flow Gas Jets [121]	30.0 %	--
Cup Burner (Coflow) [33]	--	32.5 %
Cup Burner (Coflow) [56]	--	33.0 %
Cup Burner (Coflow) [57]	27.1 %	33.0 %
Santoro (Coflow) (This Work)	20.2 %	28.0 %

Previous studies have suggested that extinguishing concentrations for cup burners depend on the fuel flow cross-sectional area. Tucker et al. reported a slight dependence on diameter when cup burner measurements were made for two burners with different diameters (2.1 cm and 0.9 cm), with the smaller diameter flame requiring lower volume fractions of extinguishing agents. [45] They attributed their observations to the smaller diameter flame having proportionally higher heat losses to the burner. Saso et al. studied the effects of burner size on cup burner measurements with liquid fuels. [135] For a n-heptane flame they found that the extinguishing volume fraction of halon 1301 decreased by roughly 30 % when the cup diameter was reduced from 3 cm to 1 cm. This dependence on burner diameter is roughly the same as suggested by the results included in Table 15. If heat losses are indeed the source of the difference, the higher conductivity of the metal fuel tube in the Santoro burner as compared to the glass cup in the cup burner will also play a role.

Smyth and Everest have reported an extinguishing volume fraction of CF_3Br of 4.1 % for propane burning on a Santoro burner. [123] They note that this value is close to the corresponding value of 4.3 % CF_3Br observed by Hamins et al for a propane flame on a cup burner. [33] The small difference between these two values is somewhat less than would be expected based on the current results for thermal agents.

3. Modeling of Extinction and Extinguishment of Methane and Propane Flames by Nitrogen

OPPDIF was used to calculate flame structures for methane and propane opposed-flow diffusion flames reacting with air containing various percentages of added nitrogen. Figure 28 and Figure 29 show the results. Several points are immediately obvious. As the percentage of added nitrogen increases, the fuel and oxidizer velocity magnitudes sufficient to cause flame extinction decrease. The maximum flame temperature at extinction also decreases with increasing nitrogen concentration. The negative slopes of the plots for T_{max} versus velocity increase in absolute value as the concentration of nitrogen increases.

Even though the results for methane and propane have similar appearances, there are quantitative differences. In Section III.D it was noted that propane flames burning in air required higher velocities, strain rates, and scalar dissipation rates to extinguish than methane flames. The maximum flame temperature at extinction was also lower for the propane flame. These trends continue as the flames are diluted with nitrogen. This can be seen in Figure 30, where values of the calculated maximum flame

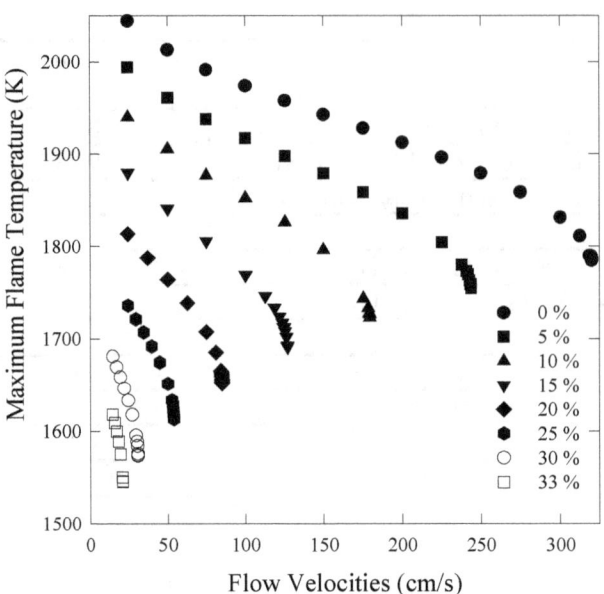

Figure 28. Maximum calculated flame temperatures are plotted against methane and oxidizer exit velocity magnitudes for a series of methane/air opposed-flow diffusion flames diluted with nitrogen. Symbols correspond to the volume fraction of added nitrogen.

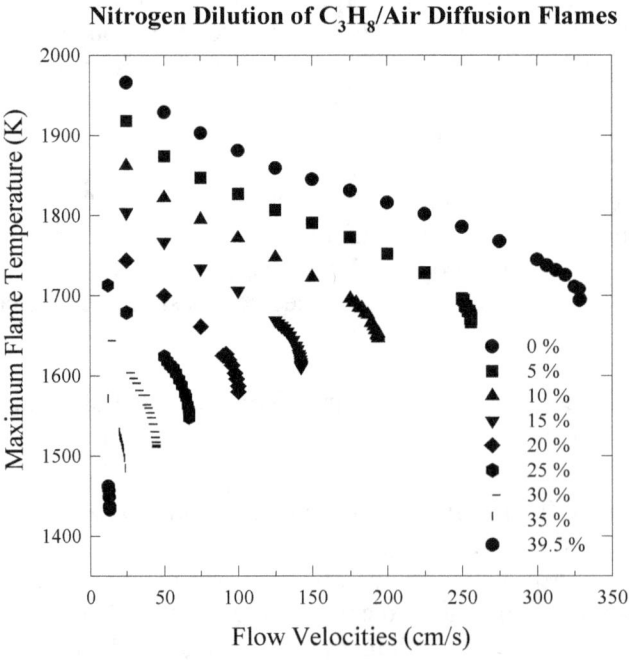

Figure 29. Maximum calculated flame temperatures are plotted against propane and oxidizer exit velocity magnitudes for a series of propane/air opposed-flow diffusion flames diluted with nitrogen. Symbols correspond to the volume fraction of added nitrogen.

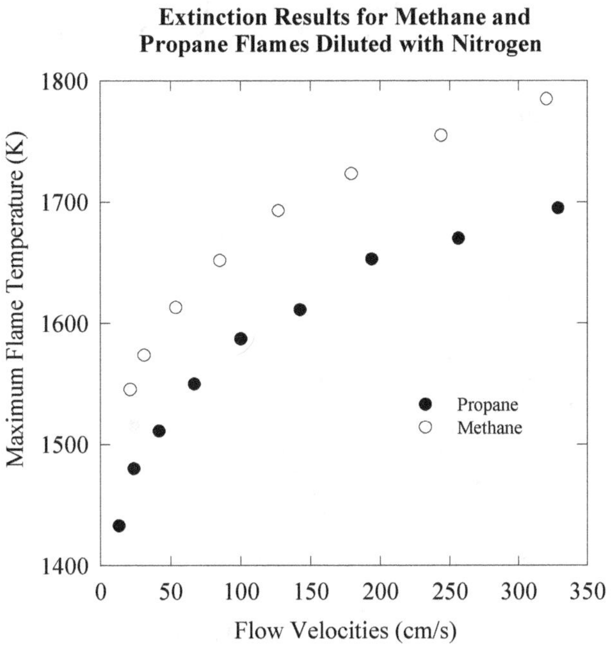

**Extinction Results for Methane and
Propane Flames Diluted with Nitrogen**

Figure 30. Values of the maximum calculated flame temperature at extinction are plotted as a function of the fuel and oxidizer flow velocities for methane and propane flames burning in air diluted by various volume fractions of nitrogen.

temperature at extinction for the two flames are plotted as a function of the fuel and oxidizer flow velocities. For lower flow velocities the differences between the maximum flame temperatures at extinction become smaller, but the values for methane remain higher for a given velocity. In Figure 31 the maximum flame temperatures have been replotted as a function of the nitrogen volume fraction. For later comparison purposes, this plot also contains adiabatic flame temperatures at stoichiometric conditions for methane and propane calculated using the program EQUIL supplied by Reaction Design. It is particularly interesting that the order of the temperatures for methane and propane is reversed when comparing the two cases. This demonstrates that either transport properties or chemical-kinetic effects play an important role in determining the flame structure. Note also that both types of temperature fall with increasing nitrogen volume fraction, but that the adiabatic flame temperature drops more quickly. As a result, the two temperatures approach each other at higher nitrogen concentrations.

The discussion thus far has been in terms of the general extinction behavior for the two flames. Now we will turn our attention to extinguishment. As discussed earlier, it is first necessary to define a characteristic extinguishing nitrogen volume fraction for the diffusion flames. As shown by the results in Table 15, experimental values depend on the diffusion flame burner configuration. It seems reasonable to use the highest experimental values, which are found for the opposed-flow porous burners. This choice is also consistent with the widespread use of this type of burner for LOI measurements. Based on the results included in Table 15, characteristic extinguishing volume fractions of 33 % and 39.5 % were chosen for methane and propane flames, respectively.

Calculated maximum flame temperatures for the two characteristic extinguishing concentrations are included in Figure 28 and Figure 29. From these, the corresponding extinguishment maximum flame temperatures can be obtained as 1545 K for methane and 1433 K for propane. These values can be compared with temperatures reported in the literature for diffusion flames near extinguishment. For methane flames experimental values of 1483 K (Tsuji burner) and 1595 K (opposed-flow burner) have

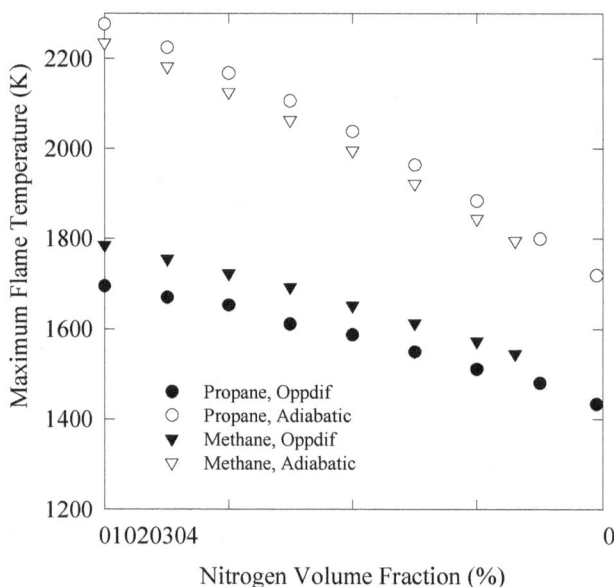

Figure 31. Values of maximum flame temperatures calculated using OPPDIF are plotted as a function of added nitrogen volume fraction for methane and propane opposed-flow flames. These values are compared with the corresponding adiabatic flame temperatures assuming stoichiometric combustion.

been reported by Ishizuka and Tsuji [48] and Puri and Seshadri [114], respectively. The agreement between the two studies is probably better than a simple comparison indicates since the thermocouple measurements of Ishizuka and Tsuji were not corrected for radiative heat losses, which would be expected to decrease the measured temperature below the actual temperature. The thermocouple measurements of Puri and Seshadri were corrected for radiative heat losses. The calculated temperature is expected to be a slight overestimate due to neglect of radiative heat losses from flame gases.

Puri and Seshadri reported that the maximum flame temperature for a propane flame at the limit condition with added nitrogen was 1656 K. This value is 77 K higher than they measured for methane. These results are in disagreement with the current calculations that predict that the propane flame will have the lower temperature for extinguishing conditions. We cannot provide an explanation for this difference. Babb et al. have reported a maximum flame temperature of 1600 K for a nitrogen-inhibited propane flame near extinction. [56] Similar measurements in a heptane flame yielded a value of 1500 K.

Based on the above discussion, we conclude that the experimental data currently available for temperatures in diffusion flames near extinction is too limited to allow a meaningful assessment of the effectiveness of the calculations in predicting flame temperatures. More careful experiments are required for such a validation.

As discussed earlier, Sheinson et al. [49] and Zegers et al. [52] have argued that hydrocarbon combustion, in general, is not sustainable when the maximum flame temperature drops below roughly 1600 K. This estimate is in reasonable agreement with the maximum flame temperatures at extinguishiment estimated by the current study, but it should be noted that it does not allow for any fuel-to-fuel variations such as those reported here.

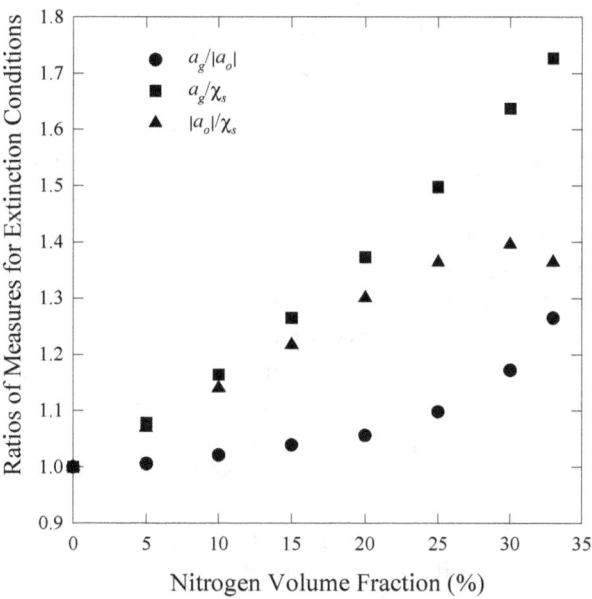

Figure 32. Ratios for the indicated measures of conditions at flame extinction are plotted as a function of nitrogen volume fraction added to the air for methane/air flames.

For the conditions corresponding to the assumed extinguishing volume fraction of nitrogen, the velocities of the fuel and air at extinction are calculated to be 21.42 cm/s for the methane flame. Corresponding values for the various measures used to characterize the flow condition at extinction are a_g = 37.5 s^{-1}, $|a_o|$ = 27.0 s^{-1}, and χ_s = 0.80 s^{-1}. It is interesting to see how the various measures vary relative to each other with added nitrogen concentration. Figure 32 shows plots of $a_g/|a_o|$, a_g/χ_s, and $|a_o|/\chi_s$ versus the percentage of added nitrogen. It is clear that the three measures have different dependencies on the amount of thermal agent added. The reason for this is the variation in the thickness of the boundary layer as the flow velocities are changed. The dependence on agent concentration requires that great care be exercised when comparing results in terms of different measures of strain rate.

The extinguishing condition for the calculated propane flame corresponds to $u_f = u_o$ = 13.4 cm/s, a_g = 29.8 s^{-1}, $|a_o|$ = 15.3 s^{-1}, and χ_s = 2.2 s^{-1}. The values for propane and methane have similar magnitudes, but are not identical. These differences are likely associated with differences in the stoichiometries and flammability for the two fuels.

An important question is: what strain rate is appropriate to use when determining the minimum value of an added thermal agent required to extinguish a buoyancy-dominated diffusion flames at normal gravity? Only limited discussions of this point are available in the literature. In their early work, Puri and Seshadri reported that extinguishment by nitrogen added to air occurred when a_g was decreased to approximately 30 s^{-1} and 35 s^{-1} for methane and propane, respectively. [114] Hamins et al. compared cup burner measurements (heptane fuel) of extinguishing concentrations for a variety of agents with corresponding measurements made in a counter-flow flame. [53] The measurements were made over a range of flow velocities characterized in terms of a global strain rate, a_{gl}, defined as

$$a_{gl} = \frac{2u_{ol}}{L_{ol}}, \qquad (16)$$

which is appropriate for an oxidizer flow impinging on a liquid surface. L_{ol} and u_{ol} are the distance between the oxidizer exit and the liquid surface and the nominal oxidizer exit velocity, respectively. In agreement with the current calculations, the strain rates necessary to induce extinction of the counterflow flame decreased with increasing concentration of added agent. When the concentrations of added agents for the opposed-flow flame were comparable to those observed in the cup burner test, the values of a_{gl} required for extinction were on the order of 50 s^{-1}. Saso et al. have reported a similar comparison. [136] For nitrogen, carbon dioxide, and a number of fluorinated hydrocarbons they found that extinction volume fractions for their opposed-flow burner agreed very well with their cup burner results when a_{gl} was equal to 60 s^{-1}. Note that Saso et al.'s definition for the global strain rate did not include the factor of 2 that appears in Eq. (16). For this reason, their results have been multiplied by this factor to allow comparison with the results of Hamins et al. They agree to within 20 %. Due to the different boundary conditions and fuels used between the current investigation and those of Hamins et al. and Saso et al., as well as slightly different definitions for the global strain rates, direct quantitative comparisons are not appropriate. However, it is clear that the strain rates have comparable magnitudes in each case. It is important to note that the strain rates appropriate for characterizing the LOI are much lower than typically used for experimental and modeling investigations of opposed-flow laminar diffusion flames.

The calculations indicate that extinguishment of diffusion flames not only occurs at a well-defined maximum flame temperature, but also with a well-defined strain rate. It is interesting to speculate as to why lower and lower strain rates are not sustained for such flames. The most likely reason is that buoyancy induced velocity results in a lower limit for the strain rate perpendicular to the flame surface. The results of Hamins et al. and the current findings suggest this minimum strain rate is on the order of a few tens of inverse seconds. [53]

F. Effects of Argon Addition on Extinction and Extinguishment of Methane Diffusion Flames Burning in Argon "Air"

As already pointed out, we have identified no additional literature measurements of LOI for thermal agents added to methane/air diffusion flames. However, Ishizuka and Tsuji did make measurements for methane burning in an artificial "air", consisting of 21 % oxygen and 79 % argon. [48] This "air" was diluted with argon until a LOI was found. The result corresponded to an added argon concentration of 54.3 %. The measured maximum flame temperature for the LOI flame was 1443 K, or roughly 40 K less than for standard air diluted with nitrogen. In order to test the ability to predict LOI concentrations of added thermal agents, a series of calculations for argon "air" diluted with argon were made.

Figure 33 shows a plot of the maximum calculated flame temperature as a function of the percentage of argon added to the argon "air" and the equal fuel and oxidizer velocity magnitudes. Comparison with Figure 28 shows that replacing nitrogen with argon has a dramatic effect on the flame behavior. First, flame temperatures are considerably higher for given diluent concentration and fuel and oxidizer velocities for the argon "air". This is due to the lower heat capacity of argon as compared to nitrogen. As a result of the increase in temperature, much higher flow velocities are required to generate the characteristic strain rate necessary to result in extinction of the methane flame. For the same reason, higher concentrations of argon must be added to achieve the LOI than in the nitrogen case.

The flame formed by oxidizer diluted with 50 % argon is calculated to undergo extinction with a maximum temperature of 1610 K and methane and oxidizer velocity magnitudes of 37.8 cm/s. The results for 54 % argon, corresponding to the experimental LOI, are 1473 K and 15.9 cm/s. These values are both slightly smaller than found for the methane/air flame diluted with nitrogen, 1545 K and 21.4 cm/s, respectively. However, they are remarkably close when one recalls that the use of argon instead of nitrogen should result in a significantly different flame structure due to changes in heat capacity and thermal diffusivity on going from nitrogen to argon. In fact, if one simply assumes that the LOI occurs

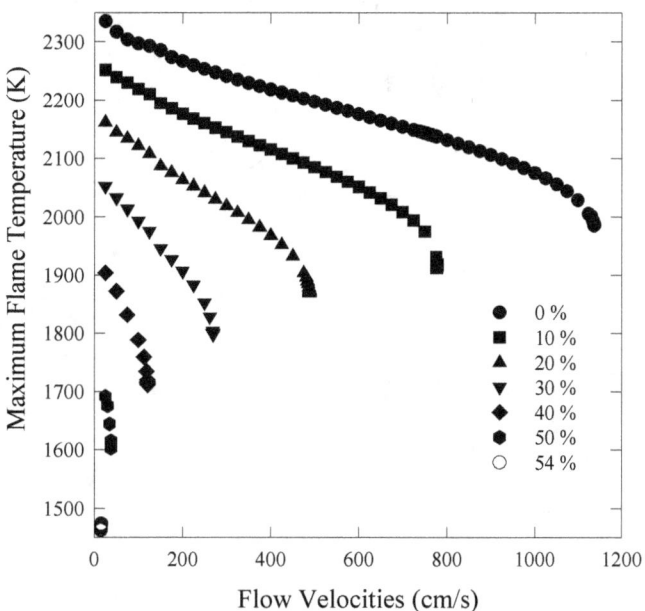

Figure 33. Maximum calculated flame temperatures are plotted against methane and oxidizer flow velocity magnitudes for a series of methane/argon "air" opposed-flow diffusion flames diluted with argon. Symbols correspond to the volume fraction of added argon.

for the same maximum flame temperature, i.e., roughly 1550 K, as for the nitrogen-diluted air flame, it is possible to estimate the required argon concentration as 52 % from Figure 33. This is only 4 % less than the experimental value. Thus, assuming that the LOI corresponds to the added concentration necessary to reduce the maximum flame temperature at extinction to 1550 K should provide excellent estimates for the amount of an arbitrary thermal agent required to extinguish a methane flame. The approach of using a limit flame temperature as the basis for estimating extinguishing concentrations extends the ideas of such researchers as Simmons and Wolfhard [41], Maĉek [42], Tucker et al. [45], Sheinson et al. [49] and Zegers et al. [52] The major difference from these earlier studies is that the effects of detailed chemical kinetics and transport properties are accounted for in estimating the extinguishing concentrations of thermal agents. In the following section this approach is used to predict extinguishing concentrations for CO_2, Ar, He, and water vapor.

G. Extinguishment of Methane and Propane Diffusion Flames by Carbon Dioxide, Argon, Helium and Water Vapor

Opposed-flow diffusion flame calculations have been used to estimate the required extinguishing concentrations, i.e., the added concentration corresponding to the LOI for nitrogen, for a number of gases that are expected to act primarily as thermal agents. Figure 34 to Figure 37 show plots of the maximum calculated flame temperature as a function of the fuel and oxidizer velocity magnitudes, where air has been diluted by the indicated percentages of argon, helium, carbon dioxide, and water vapor, respectively. It should be noted that both water and carbon dioxide are products of methane combustion and could conceivably modify the combustion chemistry slightly. However, both are believed to act primarily as thermal agents. Calculations using surrogate agents whose physical properties are identical to those for CO_2 and H_2O, but which did not participate in the flame chemistry, gave similar results, indicating that the chemical effects are indeed small.

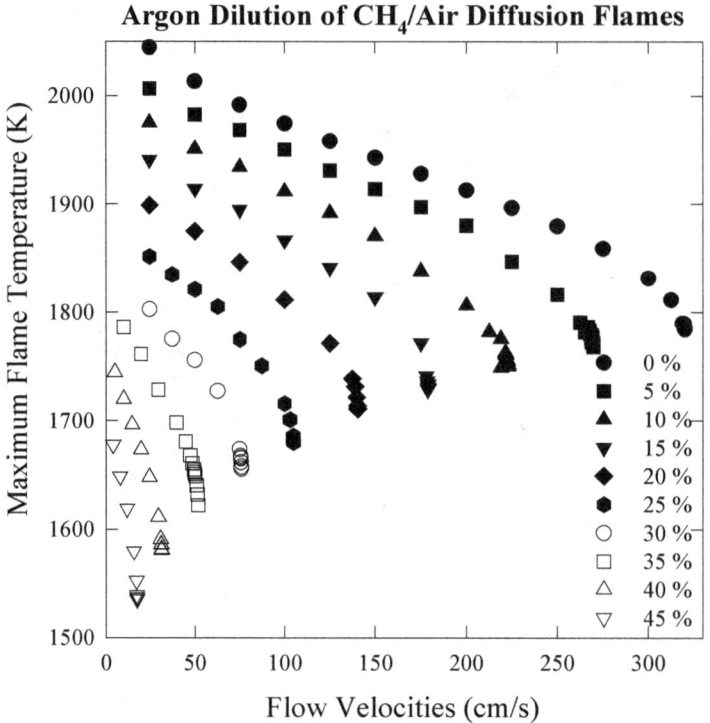

Figure 34. Maximum calculated flame temperatures are plotted against methane and oxidizer flow velocity magnitudes for a series of methane/air opposed-flow diffusion flames diluted with argon. Symbols correspond to the volume fraction of added argon.

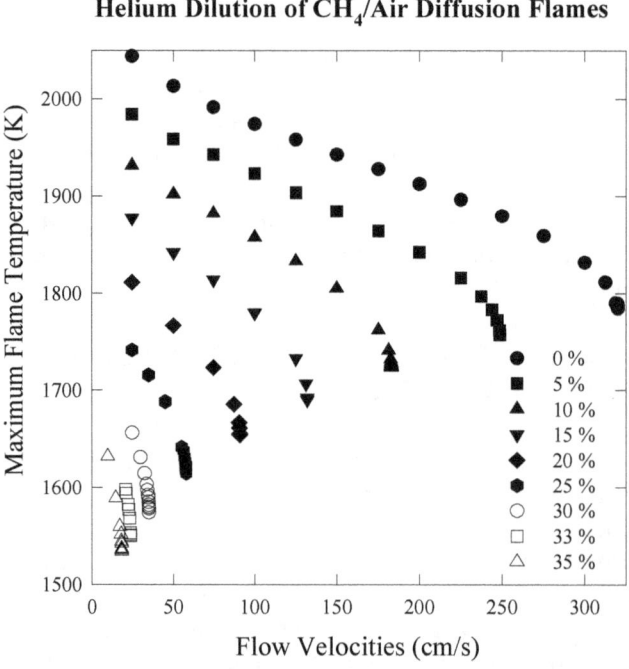

Figure 35. Maximum calculated flame temperatures are plotted against methane and oxidizer flow velocity magnitudes for a series of methane/air opposed-flow diffusion flames diluted with helium. Symbols correspond to the volume fraction of added helium.

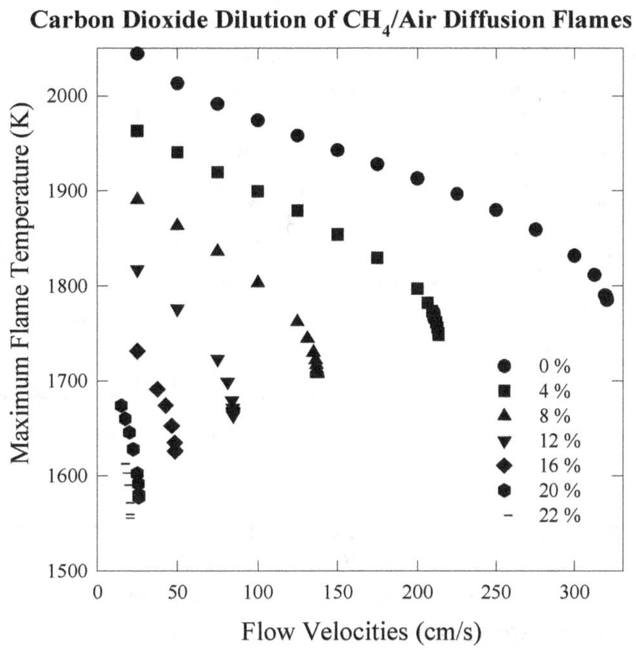

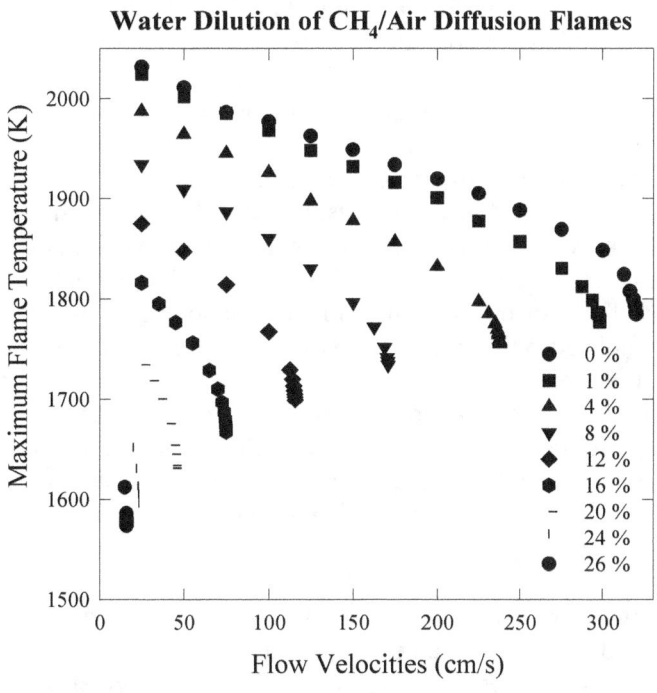

Figure 36. Maximum calculated flame temperatures are plotted against methane and oxidizer flow velocity magnitudes for a series of methane/air opposed-flow diffusion flames diluted with carbon dioxide. Symbols correspond to the volume fraction of added carbon dioxide.

Figure 37. Maximum calculated flame temperatures are plotted against methane and oxidizer flow velocity magnitudes for a series of methane/air opposed-flow diffusion flames diluted with water vapor. Symbols correspond to the volume fraction of added water vapor.

Table 16. Extinguishing Volume Fractions of Thermal Agents Calculated For Opposed-Flow Methane Flames and Measured For Heptane Flames in Cup Burner Experiments

Thermal Agent	Current Work	Cup Burner [47]	Cup Burner [49]	Cup Burner [56]	Cup Burner [33]	Cup Burner [54]	Cup Burner [55]
Nitrogen	0.33	0.302	0.30	0.33	0.32	0.30	0.336
Argon	0.43	-	0.41	-	0.41	0.38	0.433
Helium	0.33	-	0.32	-	0.31	-	-
Carbon Dioxide	0.22	0.205	0.21	0.20	0.23	0.20	0.220
Water	0.28	-	-	-	-	-	-

For water, the vapor pressure at the lower temperatures is less than required to generate the concentrations assumed in the calculations. As an example, at 300 K the water vapor pressure is 3.57 kPa, which corresponds to a volume fraction of 0.035 in atmospheric air. Therefore, for temperatures near 300 K, the higher water vapor concentrations are assumed to be supersaturated.

By assuming the limit temperature for methane flames is 1550 K, it is possible to estimate the extinguishing concentrations for each of these thermal agents. These estimates are tabulated in Table 16. As already discussed, the only experimental values for opposed-flow methane diffusion flames we have identified in the literature are for nitrogen dilution. Cup burner determinations of extinguishing concentrations using heptane as fuel have been reported for some of these thermal agents by Hirtst and Booth [47], Sheinson et al. [49], Babb et [56], Hamins et al. [33], Moore et al. [54], and Saito et al. [55]. These values are included in Table 9 for comparison purposes and are shown graphically in Figure 38.

Some variation, roughly ± 10 %, is evident in the experimental results. In general, the agreement between the experimental cup burner values for heptane and the calculated values for the methane opposed-flow diffusion flame is good. The maximum difference between values calculated for methane and the experimental values for heptane is 12 %, with the vast majority being less than 10 %. The degree of agreement between the current results and the measurements of Saito et al. is particularly good. [55] With the exception of carbon dioxide, the cup burner results are somewhat lower than predicted for the opposed-flow flame. The differences between the cup burner and opposed-flow flame results could be due to the use of different fuels or to differences in burner configuration. The heptane fires are burning just above a liquid fuel in a coflow of oxidizer, and it may be easier to blow out this type of flame than the opposed-flow diffusion flame that is stabilized away from surfaces. It has also been shown that extinguishing concentrations for coflow flames depend on the burner diameter. The close tracking of the calculated results and the experimental findings suggests that detailed chemical-kinetic modeling can accurately predict the amount of a thermal agent required to extinguish opposed-flow diffusion and cup burner flames.

The differences in the extinguishing concentrations of helium and argon are interesting since these agents are both monatomic gases and have the same heat capacities. The difference is clear in both the calculations and the experimental values. The fact that helium is a more efficient extinguishing agent means that at least one other parameter, in addition to heat capacity, is important in determining extinguishing efficiency. A related observation was reported by Coward and Hartwell for the inerting of premixed flames and was attributed to the much higher thermal conductivity of helium, which distributes the heat of combustion over a larger region of space and therefore weakens the flame. [137] The same explanation is most likely valid for diffusion flames. Sheinson et al. reached the same conclusion. [49]

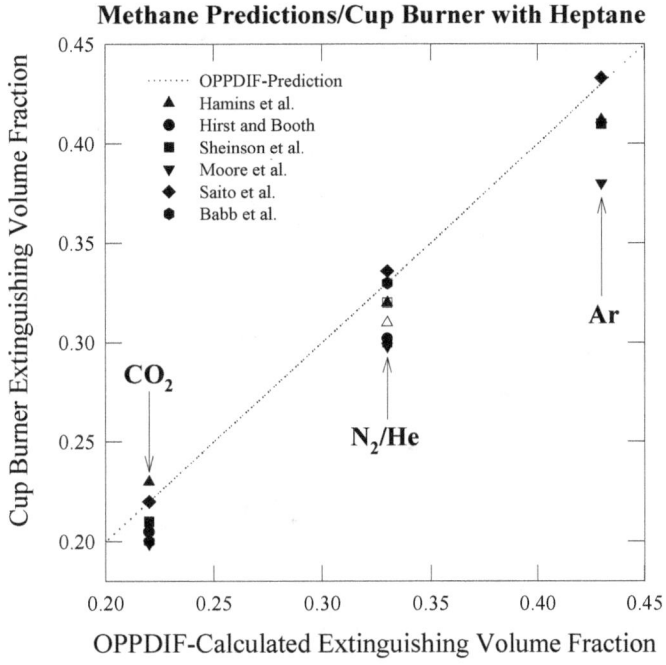

Figure 38. Detailed chemical-kinetic modeling predictions for extinguishing volume fractions of nitrogen, carbon dioxide, argon, and helium for methane flames are compared with various experimental cup burner measurements using n-heptane. Model values for helium and nitrogen overlap and open symbols are used to identify the helium results. References for the cup burner measurements are Hamins et al. [33], Hirst and Booth [47], Sheinson et al. [49], Moore et al. [54], Saito et al. [55], and Babb et al. [56]

In Figure 39 values of the calculated maximum flame temperature at extinction are plotted as a function of agent volume fraction for the five thermal agents. For each case extinguishment is assumed to occur when the calculated maximum flame temperature drops to 1550 K. Corresponding values for the adiabatic flame temperature are also included in the figure. It can be seen that the adiabatic flame temperatures for N_2, CO_2, Ar, and H_2O in air at the extinguishing concentration fall relatively close together (roughly 1800 K), with that for CO_2 being somewhat higher. For helium the calculated adiabatic flame temperature at the extinguishing condition is more than 100 K higher (note that the adiabatic flame temperatures for given added volume fractions of helium and argon are the same since these two gases have identical heat capacities). The calculated maximum flame temperatures for helium and nitrogen for a given volume fraction fall very close together, even though nitrogen has a higher heat capacity. These points emphasize the enhanced effectiveness of helium as compared to the other thermal agents.

Inspection of the various curves for the dependence of calculated maximum flame temperature on the fuel and oxidizer velocity magnitudes in Figure 28 and Figure 34 - Figure 37 shows that for given velocities and extinction temperatures, they have very similar shapes. This is true despite the fact that the amount of added agent and, therefore, the oxygen concentration and stoichiometric ratio vary dramatically with agent. This is confirmed by the results plotted in Figure 40, which show the maximum flame temperature at extinction as a function of the exit velocity magnitudes for the five thermal agents added to air as well as the results for argon added to argon "air". The data fall within a narrow band on either side of a well-defined curve.

Closer inspection of Figure 40 shows that the data for argon "air" tend to fall at higher temperatures by amounts that decrease with dilution, while the water vapor data tend to lie near the top of the band by

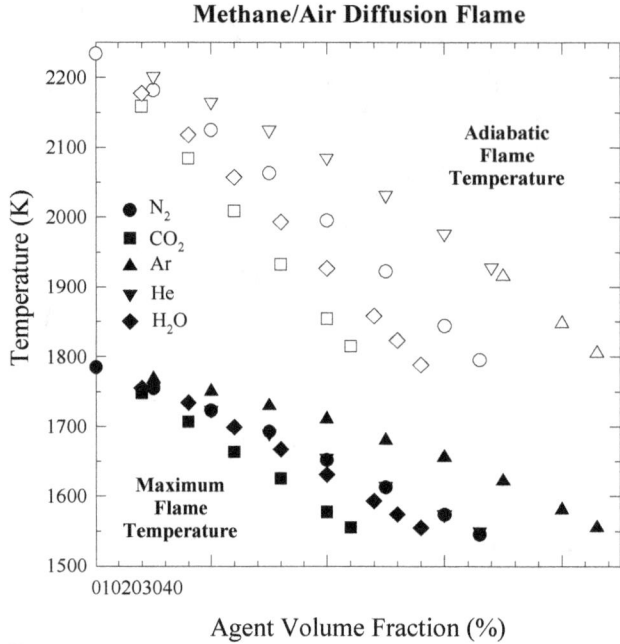

Figure 39. Values of maximum flame temperatures (solid symbols) calculated using OPPDIF and the corresponding adiabatic flame temperatures (open symbols) calculated using EQUIL for methane flames are plotted as a function of added agent volume fraction in air for five thermal agents. Note that the adiabatic flame temperatures for argon and helium overlap and that the maximum calculated flame temperatures for helium and nitrogen fall very close together.

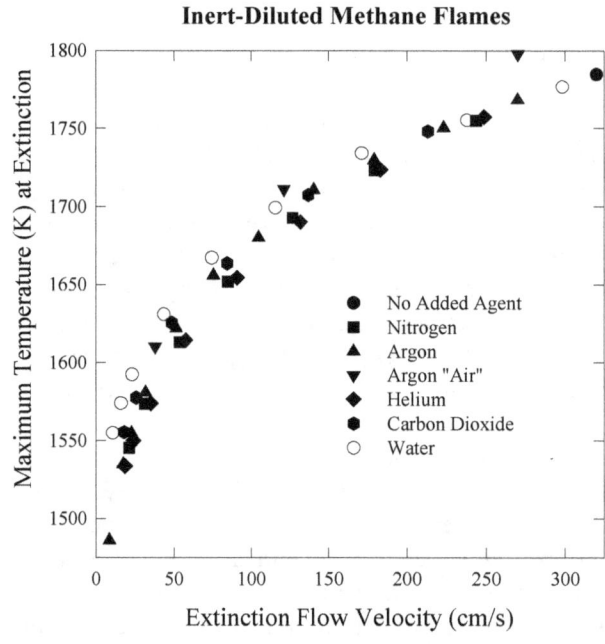

Figure 40. Values of maximum flame temperature at extinction are plotted as a function of the corresponding fuel and oxidizer velocity magnitudes for a range of volume fractions of the various thermal agents.

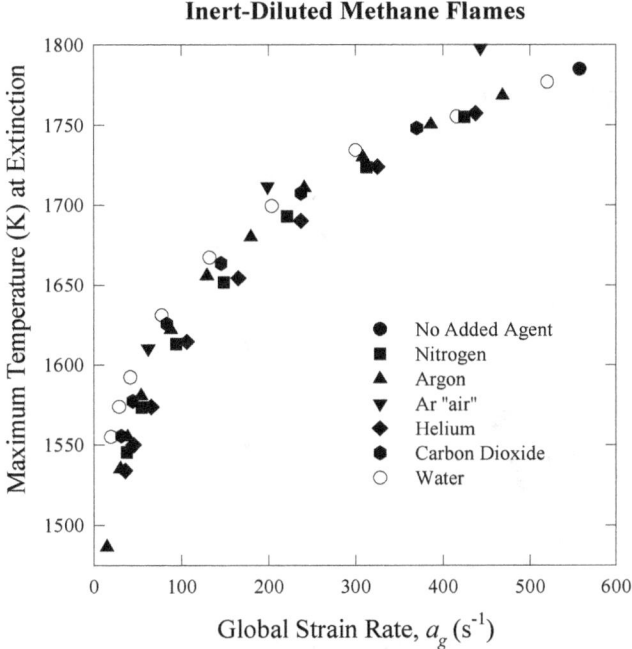

Figure 41. Values of maximum flame temperature at extinction are plotted as a function of the corresponding global strain rates, a_g, for a range of volume fractions of the various thermal agents.

amounts that increase with dilution. The difference between adding argon to air and argon "air" is most likely due to the different physical properties of argon and nitrogen. Apparently, these differences are most important for high strain rates and are relatively minor at the lower strain rates typical of buoyancy-dominated diffusion flames. Recall that the estimated value of the extinguishing concentration for argon added to argon "air" was in good agreement with the experimental value of Ishizuka and Tsuji. [48] The increasing differences with decreasing strain rate for water may be due to two possible effects. The first is the higher thermal diffusivity expected for the relatively small water molecule. However, the results for helium, which has a much higher thermal diffusivity, fall close to the results for nitrogen, argon, and carbon dioxide. The second possibility is that the water vapor slightly modifies the chemical structure of the flame and seems the more likely of the two. Increases in water concentration can interact with the flame chemistry through a number of reactions including the well-known water-gas shift reaction,

$$H_2O + CO \Leftrightarrow CO_2 + H_2. \qquad (17)$$

The addition of water to the oxidizer is expected to drive this equilibrium towards the right, resulting in additional heat release and a slightly higher temperature, as observed. The findings of Lentati and Chelliah indicate that water does display a small chemical effect. [74] Dilution with carbon dioxide would be expected to drive the equilibrium in Reaction 17 in the opposite direction, however, this effect on extinguishment must be smaller since the carbon dioxide results lie very close to those for nitrogen, argon, and helium.

Simply plotting the results in terms of the exit velocity magnitudes of methane and oxidizer at extinction leads to a collapse of the results for different thermal agents to a well-defined curve. It is of interest to see if the collapse is improved by plotting the data in terms of the global strain rate given by Eq. (14), a_g, or the maximum absolute value of the characteristic strain rate element on the oxidizer side lying outside of the boundary layer, $|a_o|$. Figure 41 and Figure 42 show the data plotted in terms of these two

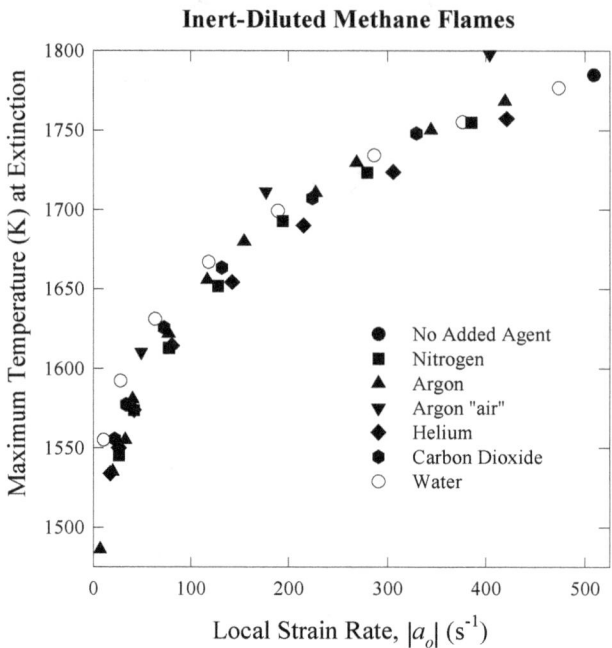

Figure 42. Values of maximum flame temperature at extinction are plotted against the corresponding characteristic local strain rates, $|a_o|$, defined as the highest absolute value of strain rate observed outside of the boundary layer on the oxidizer side of the flame, for a range of volume fractions of the various thermal agents.

characteristic strain rates. The overall appearances of the two plots are very similar to Figure 40, which is simply in terms of the exit velocities. For this reason, further discussion is based only on the flow velocity magnitudes for methane and oxidizer.

A more quantitative idea of the degree of correlation between the maximum calculated flame temperatures at extinction and the fuel and oxidizer flow velocity magnitudes is obtained from Figure 43 where the results for N_2, Ar, He, and CO_2 addition have been consolidated and fit to two piecewise fourth-order polynomials on either side of 105 cm/s. The resulting fits are drawn as a solid line. The coefficients of determination for the nonlinear curve fits were 0.97 and 0.98, which indicate a high degree of confidence in the fits. The polynomials can be used to predict values of maximum temperature for given exit velocities.

To determine if there is a relationship between added agent volume fraction and the calculated maximum temperature at extinction, the temperatures were plotted versus the ratio of the volume fraction of added agent and that required for extinguishment as shown in Figure 44. The results for argon added to argon "air" lie well away from the remaining data. The trend of less effectiveness of water vapor at low velocities is also evident. While the data partially collapse for the remaining agents, the spread indicates that simply calculating a single extinction condition for a single agent volume fraction will not suffice for predicting the extinguishing volume fraction for that agent.

Due to the long run times required for calculating opposed-flow propane diffusion flames, extinction curves for a wide range of thermal agent concentrations were not performed. Instead, more limited calculations designed to determine the extinguishing concentration, defined as the volume fraction having a maximum flame temperature of 1433 K at extinction, were performed. Figure 45 to Figure 47 show the results with CO_2, Ar, and He added to the air. The corresponding extinguishing volume fractions are

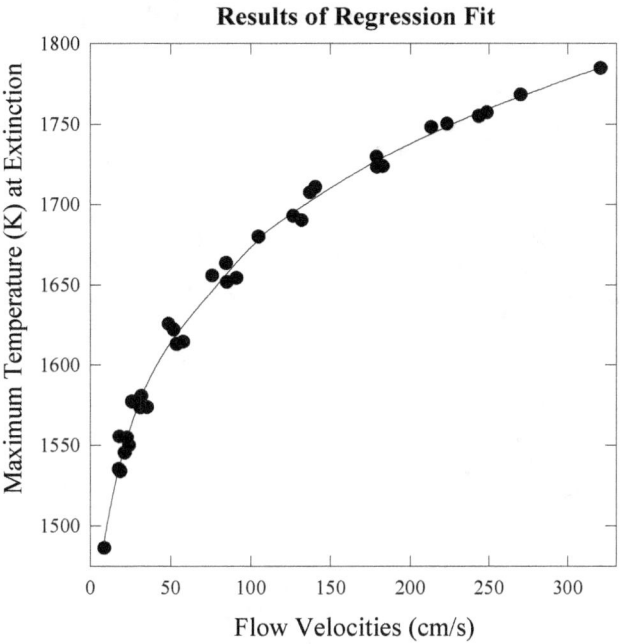

Figure 43. Maximum flame temperatures at extinction for N_2, Ar, He, and CO_2 as a function of the corresponding fuel and oxidizer velocity magnitudes are fit by two piecewise fourth-order polynomials (solid line).

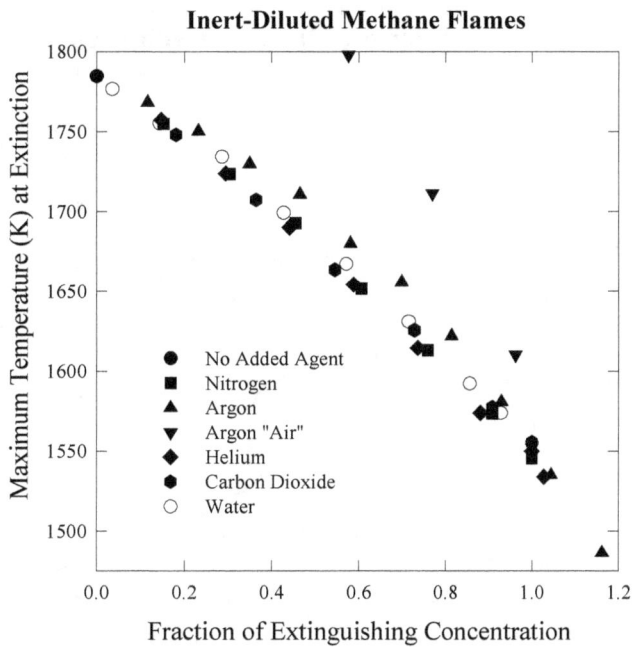

Figure 44. The maximum temperature at extinction for opposed-flow methane flames is plotted as a function of the volume fraction of a thermal agent added to air divided by the volume fraction calculated to be required for flame extinguishment.

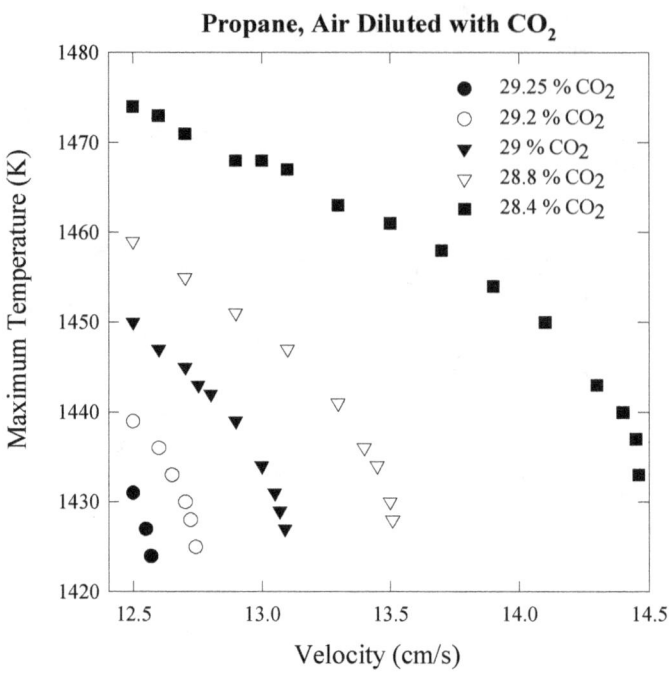

Figure 45. Calculated maximum flame temperatures are plotted as a function of velocity for a propane flame burning in air diluted with the indicated volume fractions of carbon dioxide. The extinguishing volume fraction is defined to be that (28.4 %) which yields an extinction temperature of 1433 K.

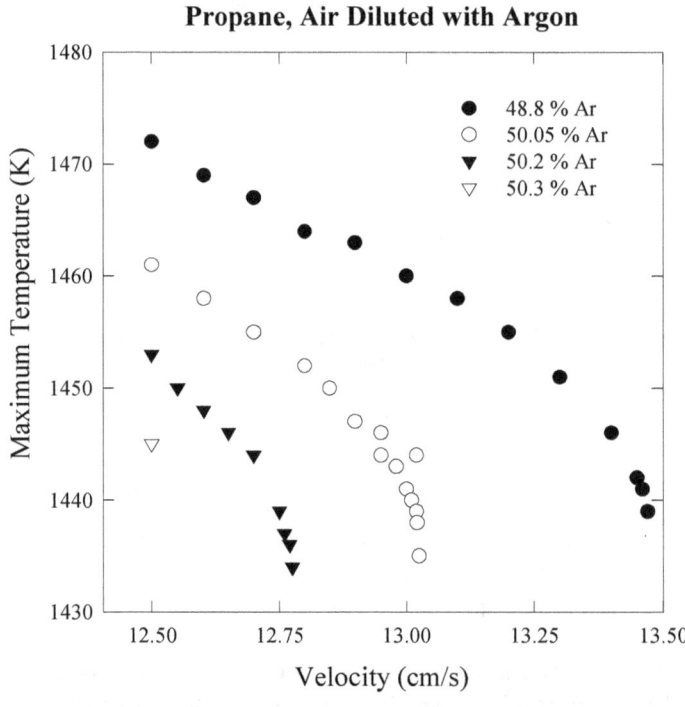

Figure 46. Calculated maximum flame temperatures are plotted as a function of velocity for a propane flame burning in air diluted with the indicated volume fractions of argon. The extinguishing volume fraction is defined to be that (50.2 %) which yields an extinction temperature of 1433 K.

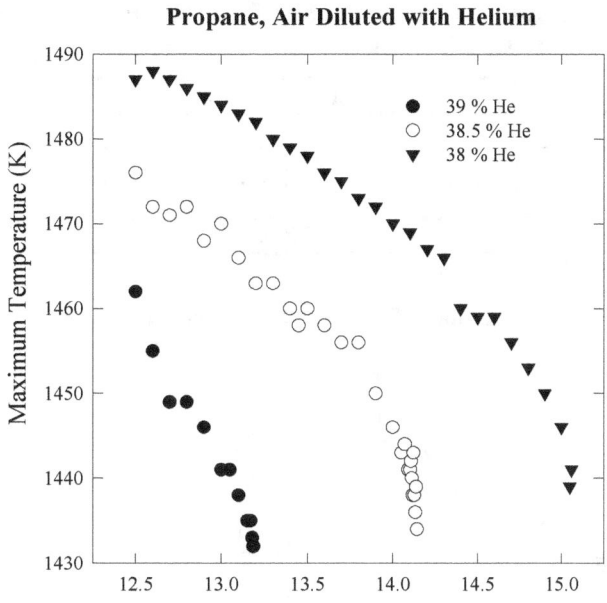

Figure 47. Calculated maximum flame temperatures are plotted as a function of velocity for a propane flame burning in air diluted with the indicated volume fractions of helium. The extinguishing volume fraction is defined to be that (39 %) which yields an extinction temperature of 1433 K.

Table 17. Calculated Extinguishing Volume Fractions for Methane and Propane Opposed-Flow Diffusion Flames for the Indicated Thermal Agents

Thermal Agent	Methane Flame	Propane Flame
Nitrogen	33 %	39.5 %
Carbon Dioxide	22 %	28.4 %
Argon	43 %	50.2 %
Helium	33 %	39.0 %

28.4 %, 50.2 %, and 39 %, respectively. These values are summarized in Table 17 along with the corresponding results for the methane diffusion flame. The relative effectiveness of the agents is the same for both fuels even though a considerably higher volume fraction of a given agent is required to extinguish the propane flame than the methane flame. It is particularly noteworthy that the required concentrations of nitrogen and helium are nearly identical for both flames. This suggests that while fuel effects influence the volume fraction of an agent required for extinguishment, such effects do not have a large influence on the relative effectiveness of different thermal agents.

Premixed adiabatic flame temperatures have been calculated for stoichiometric mixtures of propane and air diluted with extinguishing volume fractions of the thermal agents. The results are summarized in Table 18 along with the corresponding results for methane flames. The results for air diluted with N_2, CO_2, and Ar are remarkably close. As found for methane, the calculated adiabatic flame temperature for air diluted with helium is much higher than for the other thermal agents. The differences between methane and propane flame indicated by the results in Table 18 follow the same trend as identified in Figure 31, calculated adiabatic flame temperatures for extinguishing volume fractions of a given agent are lower for propane flames than for methane flames. This is true even though, for a given amount of air

Table 18. Adiabatic Flame Temperatures Calculated for Methane and Propane Premixed Flames Burning in Air Diluted with Extinguishing Volume Fractions of Thermal Agents

Thermal Agent	Methane Flame	Propane Flame
Nitrogen	1796 K	1719 K
Carbon Dioxide	1815 K	1722 K
Argon	1805 K	1724 K
Helium	1928 K	1903 K

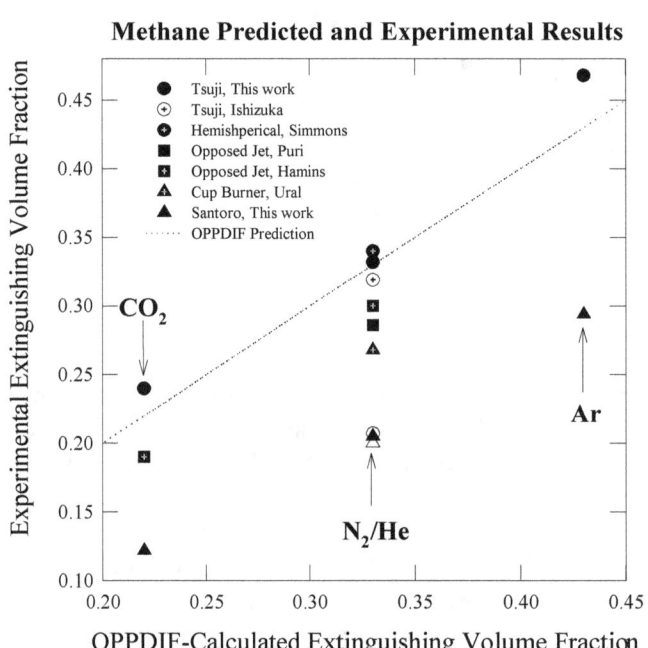

Figure 48. Experimental values of extinguishing volume fractions for N_2, CO_2, Ar, and He for counterflow porous (circles), opposed-jet (squares) and coflow (triangles) burners are plotted against the corresponding OPPDIF predictions (dotted line) for methane-air diffusion flames. Experimental values are taken from the present work as well from the following references: Ishizuka and Tsuji [48], Simmons and Wolfard [41], Puri and Seshadri [114], Hamins [121], and Ural [57]. Current results for helium are indicated by open symbols.

dilution, propane flames have higher adiabatic flame temperatures. These results support the conclusion that propane flames are more difficult to extinguish.

It is of interest to compare the predictions of the calculated extinguishing concentrations with experimental data for extinguishment of diffusion flames. Figure 48 and Figure 49 show plots of experimental values of extinguishing volume fractions for N_2, CO_2, Ar, and He versus the values calculated using OPPDIF. The dotted lines on the graphs correspond to the predicted results, which are also included on the plot.

The strong dependence of the experimental results on burner type discussed earlier is evident in both plots. In particular, the values for the Santoro burner measured during the current investigation are considerably lower for both fuels. These results do fall on a straight line indicating that the calculations have adequately captured the qualitative dependence on extinguishment for this flame type. Close inspection shows that the slopes of lines that would fit the Santoro-burner results are slightly less than that obtained from the OPPDIF predictions. It is particularly noteworthy that both calculations and

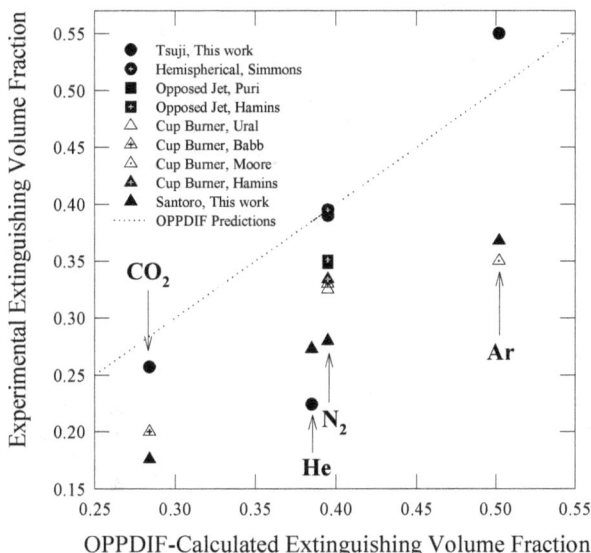

Figure 49. Experimental values of extinguishing concentrations for N_2, CO_2, Ar, and He for counterflow porous (circles), opposed-jet (squares) and coflow (triangles) burners are plotted against the corresponding OPPDIF predictions (dotted line) for propane-air diffusion flames. Experimental values are taken from the present work as well from the following references: Simmons and Wolfhard [41], Puri and Seshadri [114], Hamins [121], Ural [57], Babb et al. [56], Moore et al. [54], and Hamins et al. [33].

experiments indicate that nitrogen and helium have very similar extinguishing concentrations. The same was shown to be true for cup burner measurements with heptane fuel (see Figure 38 and Table 16).

Inspection of Figure 48 and Figure 49 shows that experimental extinguishing volume fractions for argon and carbon dioxide measured using the DLAFSS are in reasonable agreement with the model predictions. Recall that the extinguishing concentrations for nitrogen are tied to experimental values for porous burners in opposed-flow configurations, so this is agreement is to be expected. Unfortunately, the experimental results for helium lie well below the predicted extinguishing volume fractions for both fuels. As discussed earlier, it is likely that the strain rates present during the helium measurements were higher than required for buoyancy-dominated burning due to limitations of the flow system. These higher strain rates are expected to reduce the amount of thermal agent required for extinguishment as observed.

Experimental values from the literature are limited. Only two cases are available where measurements for the same burner have been reported for more than one agent. For methane, Hamins has reported extinguishing concentrations for nitrogen and carbon dioxide in an opposed-jet burner. [121] The two results fall on a line that lies below, but parallels the calculated values. Similar comments apply to the cup burner measurements of Babb et al. for a propane flame. [56]

Even though the available experimental data for comparison are limited, extinguishing concentrations for diffusion flames vary with burner type, and there are some inconsistent data, the results summarized here suggest detailed chemical-kinetic modeling combined with the concept of a limit temperature does a good job of capturing the relative effectiveness of thermal agents for extinguishing diffusion flames as well as the effects of using different fuels.

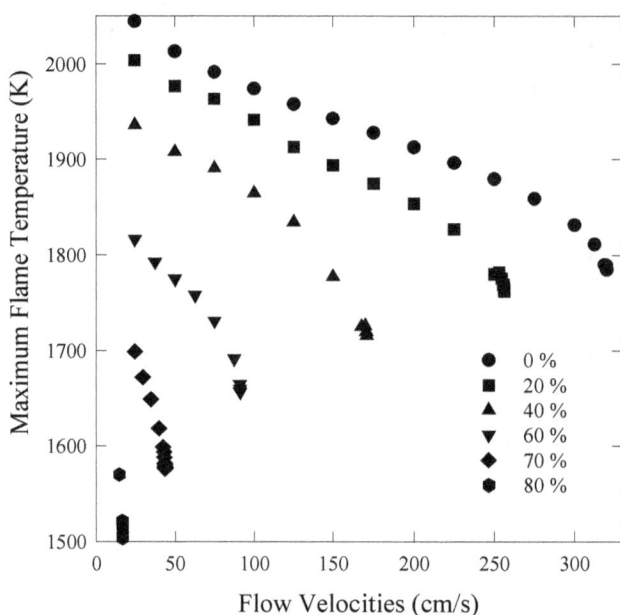

Figure 50. Maximum calculated flame temperatures are plotted against methane and oxidizer exit velocity magnitudes for a series of methane/air opposed-flow diffusion flames diluted with a surrogate agent similar to argon, but with $C_p = 0$. Symbols correspond to the volume fraction of surrogate agent added to the air.

H. Effects of Dilution and Heat Extraction Location Relative to the Flame Front on Extinction and Extinguishment of Diffusion Flames

An important advantage of modeling investigations is the ability to perform calculations for conditions that are not physically possible in order to learn details concerning the role of various system parameters. In this section this approach is applied to methane opposed-flow diffusion flames. One question that has been the subject of speculation is the relative importance of heat extraction and dilution on the effectiveness of a thermal agent. Here dilution effects refer to any role of a thermal agent that is not simply due its ability to physically lower the flame-zone temperature. In order to obtain insights into this behavior, an artificial agent was created starting with argon and setting its heat capacity to zero. Direct comparison with the results for added argon allows the relative roles of heat removal and dilution to be characterized. Sheinson et al. have discussed the effects of dilution on extinguishment. [49] They concluded that they are relatively small compared to direct heat removal due to heat capacity for the thermal agents CF_4 and SF_6. Zegers et al. have suggested that dilution effects play a role for relatively inefficient thermal agents such as nitrogen. [52]

Figure 50 is a plot of calculated maximum flame temperature versus fuel and oxidizer velocity magnitudes for the zero-heat-capacity argon added to air. It is obvious that this species does decrease the strength of the flame, but that its effect is much smaller than for the argon results shown in Figure 34. Based on an extinguishment temperature of 1550 K, the extinguishing concentration is estimated as 73 %, or roughly 1.7 times greater than required for argon. This corresponds to an oxygen concentration of 5.7 % at the LOI. Interestingly, the dependence of the maximum flame temperature at extinction on the fuel and oxidizer velocities is very similar for both sets of calculations as can be seen in Figure 51, where they are compared. The solid line is that shown in Figure 43, which was derived by fitting to the methane results for N_2, Ar, He, and CO_2. The data for the zero-heat-capacity argon fall slightly below those for

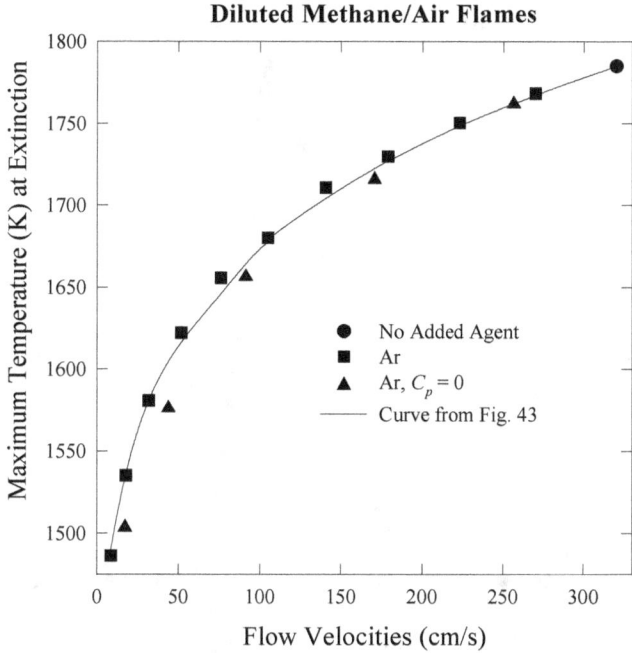

Figure 51. Values of maximum flame temperature at extinction are plotted as a function of fuel and oxidizer extinction velocity magnitudes for a range of concentrations of argon and a similar surrogate agent having $C_p = 0$. The solid line is the result of the regression fits shown in Figure 43.

argon, but it is clear that simple dilution of air has a similar effect on the flame as heat absorption by an inert. Assuming that the effects of heating an inert and dilution are additive, as suggested by this close correspondence, the effectiveness of Ar as a thermal agent is estimated to be 41 % due to dilution and 59 % due to heat extraction.

The availability of detailed flame structure information from the calculations allows the effect of a zero-heat-capacity diluent to be understood. As a starting point, it is important to recognize that flames burning in air and air diluted with a zero-heat-capacity agent have identical adiabatic flame temperatures. This suggests that the observed decreases in maximum temperature must be due to redistribution of the heat generated by combustion over a larger region of space and/or a reduction in the amount of heat released. In order to investigate this point, results for a methane counterflow diffusion flame burning in air are compared with those for a flame burning in air diluted with 60 % of the zero-heat capacity diluent in Figure 52. It is evident that while the maximum flame temperature is significantly reduced by the addition of the agent, the spatial temperature distributions are very similar for the two flames. This suggests that the primary effect of dilution is to reduce the amount of heat released.

Figure 53 shows CH_4 and O_2 volume fractions as a function of distance from the burner on a semi-log plot for both flames. It can be seen that while the primary fuel is completely reacted at the flame surface, some O_2 "leaks" from the oxidizer side to the fuel side. O_2 reaching the fuel side of the reaction layer eventually diffuses to a location where the temperature is low enough that reaction can no longer occur. At this point the O_2 acts like a diluent having a heat capacity with the result that the maximum temperature in the flame zone is reduced. Close inspection of Figure 53 shows that a significantly larger fraction of the available O_2 leaks through to the fuel side of the flame when the oxidizer is diluted air as opposed to air. This difference explains the lower maximum flame temperatures calculated for burning in air diluted by an agent having no heat capacity.

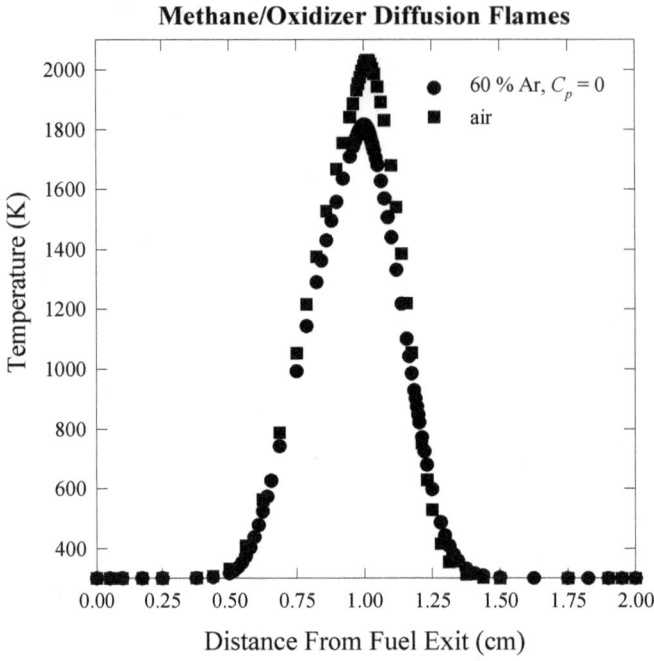

Figure 52. Temperature is plotted as a function of distance from the fuel exit for methane flames burning in air and air mixed with 60 % of a diluent having similar properties to Ar, but with zero heat capacity. Fuel and air exit velocity magnitudes are 25 cm/s for both flames.

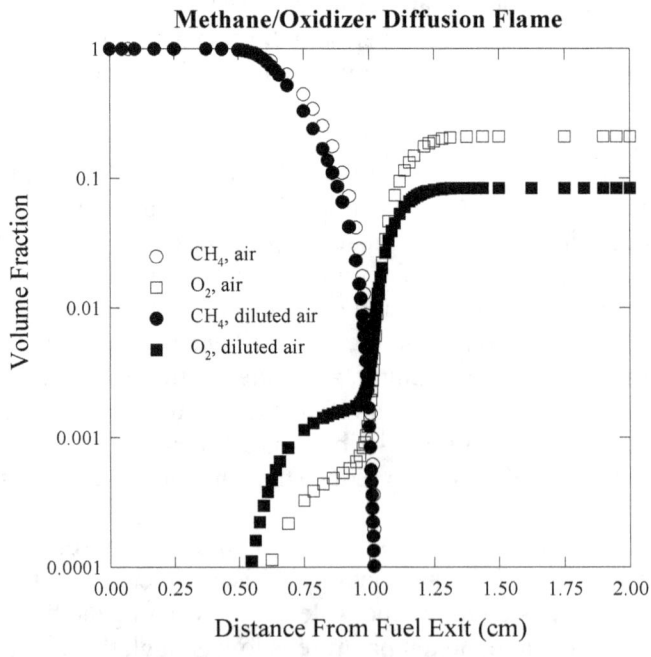

Figure 53. Calculated methane and oxygen volume fractions are shown on a semi-log plot as a function of distance from the fuel exit for methane flames burning in air (open symbols) and air mixed with 60 % of a diluent having similar properties to Ar, but with zero heat capacity. Fuel and air exit velocity magnitudes are 25 cm/s for both flames.

The leakage of O_2 through the flame front is due to the finite rates of reaction for oxidation reactions. Liñán and Williams provide a good introductory discussion of this behavior and the reasons for it. [138] Oxygen leakage was observed experimentally several decades ago [139] and was predicted by early calculations of counter-flow diffusion flames [88]. Du and Axelbaum discuss the effect of O_2 leakage with respect to a series of methane flames where they systematically varied the stoichiometric mixture fraction while maintaining a constant adiabatic flame temperature. [117] These authors found that the flame strength increased with increasing stoichiometric mixture fraction and argued that this was due to the shift of the O_2 profile into regions of higher temperature. Decreased O_2 flame leakage was also evident in the calculations with higher stoichiometric mixture fraction. Lentati and Chelliah have also noted the importance of oxygen leakage for understanding flame extinction. [74] The increased O_2 leakage and decreased maximum flame temperature observed with decreasing stoichiometric mixture fraction found for the current calculations are consistent with these trends.

One of the goals of the current work was to test the hypothesis that the effectiveness of a thermal agent depends on the location, relative to the high temperature flame zone, where heat extraction occurs. A surrogate thermal agent, X, was used for these calculations. The molecular weight, thermodynamic properties, and transport properties of X are identical to those of argon, but it is capable of undergoing a simple reaction in the presence of ambient bath gases to generate a new species, Y, i.e.,

$$X + M \rightarrow Y + M. \qquad (18)$$

Y is also very similar to argon, the only difference being that its heat of formation is assigned an arbitrary positive value instead of being zero. As a result, when Reaction (18) takes place it extracts heat and cools the local surroundings by an amount equal to the heat of reaction, $\Delta H_{X \rightarrow Y}$. Since X and Y do not react with any other species, the reaction is effectively simply a heat sink, and X therefore meets the definition of a thermal agent.

The rate constant for Reaction (18) is expressed as

$$k_{X \rightarrow Y} = AT^\beta e^{-E_a/RT}, \qquad (19)$$

where A is the pre-exponential factor, β is the temperature exponent, E_a is the energy of activation, R is the gas constant, and T is temperature. By varying the parameters A, β, and E_a it is possible to change the temperature range over which the reaction occurs and hence the location relative to the flame zone for heat extraction. For the calculations which follow, initial values were chosen for A and β, and only the value of E_a was changed in order to vary $k_{X \rightarrow Y}$. One problem faced in such calculations is that due to the variations in the residence time, which depends on the velocities of the fuel and oxidizer, the completeness of reaction over a given temperature range can vary with initial conditions. For this reason, a set of calculations was performed for one value of fuel and oxidizer exit velocity magnitudes in which E_a was varied. A relatively low value of 25 cm/s was chosen because this condition is close to that for which flame extinguishment occurs. An initial calculation was run for which the parameters in Eq. (19) were set to $A = 1 \times 10^{10}$ cm^3/(mol⋅s), $\beta = 0$, and $E_a = 25.1$ kJ/mol, and the heat of formation for Y was chosen to be 96.1 kJ/mol. The concentration of X added to the air was 5 %. Figure 54 compares the resulting temperature profile across the flame with that calculated for the case when 5 % argon is added. The heat extracted by agent X has lowered the temperature on the oxidizer side below ambient. The maximum flame temperature with X is reduced, and its position has been shifted towards the oxidizer side relative to the case with argon.

It is clear from Figure 54 that the majority of the heat extraction is occurring at low temperatures for $E_a = 25.1$ kJ/mol. This is evident in Figure 55 which shows X and Y volume fractions plotted as a function of

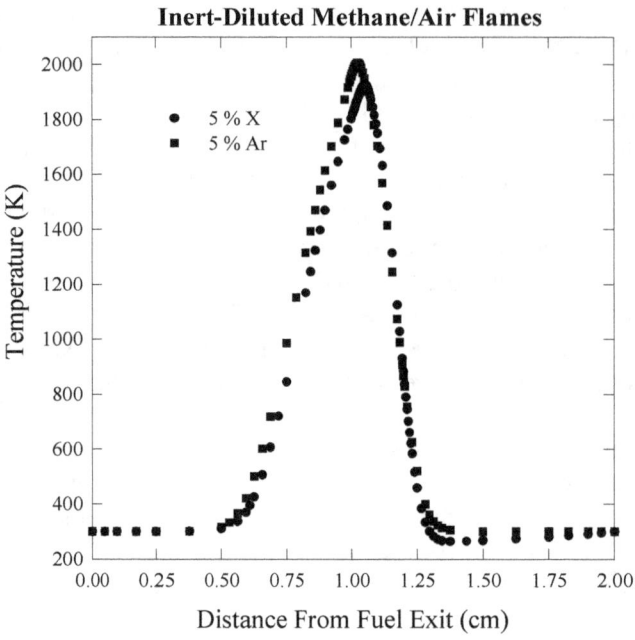

Figure 54. Calculated flame temperature is plotted as a function of distance from the fuel exit for methane/air diffusion flames where the oxidizer is diluted with 5 % of either argon or the hypothetical agent X which reacts to form Y with E_a = 25.1 kJ/mol and $\Delta H_{X \rightarrow Y}$ = 96.1 kJ/mol. Methane and oxidizer exit velocity magnitudes are 25 cm/s.

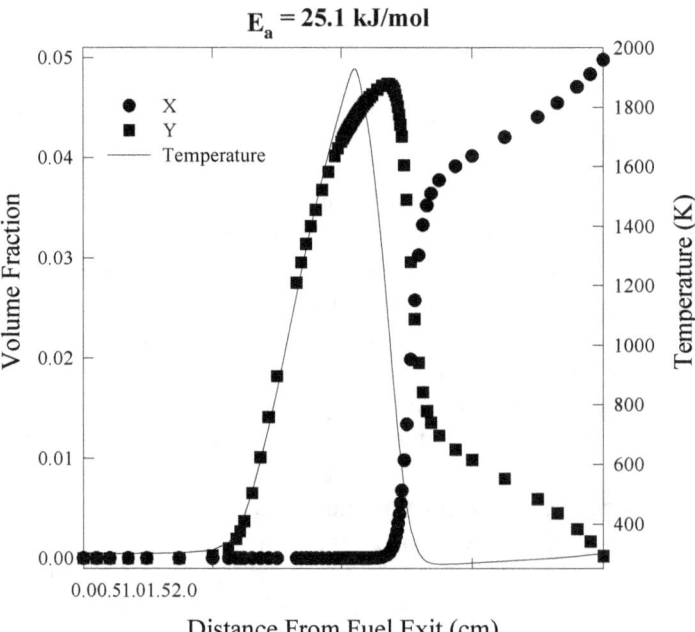

Figure 55. The volume fractions of X and Y are plotted as a function of distance from the fuel exit for a methane flame burning in air with 5 % added X. X reacts to form Y with E_a = 25.1 kJ/mol and $\Delta H_{X \rightarrow Y}$ = 96.1 kJ/mol. The spatial variation of temperature is also shown as the solid line. Methane and oxidizer exit velocity magnitudes are 25 cm/s.

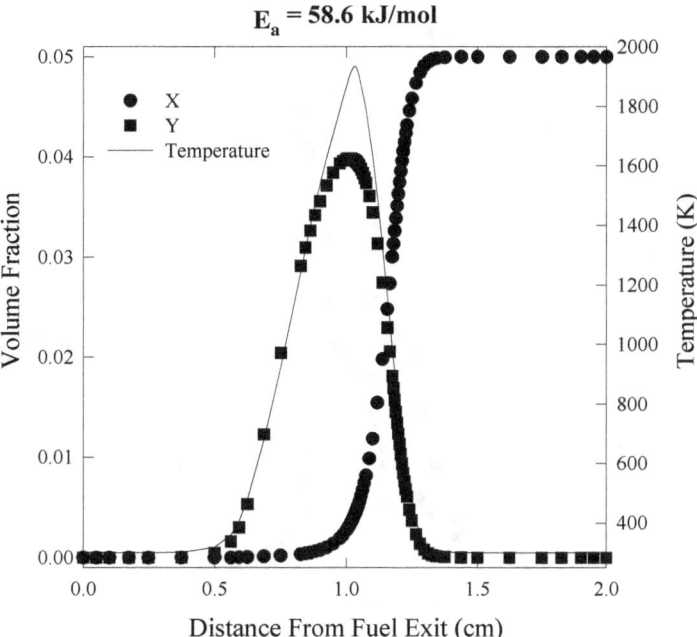

Figure 56. The volume fractions of X and Y are plotted as a function of distance from the fuel exit for a methane flame burning in air with 5 % added X. X reacts to form Y with $E_a = 58.6$ kJ/mol and $\Delta H_{X \to Y} = 96.1$ kJ/mol. The spatial variation of temperature is also shown as the solid line. Methane and oxidizer exit velocity magnitudes are 25 cm/s.

distance from the fuel exit. The corresponding temperature profile is included as the solid line. The conversion of X to Y is well advanced before the boundary layer is reached at roughly 1.3 cm from the fuel exit, and then rapidly accelerates. The concentration of X has become nearly zero by the time the location on the oxidizer side is reached where the temperature has increased to 1200 K. The maximum calculated temperature for this flame is 1928 K, which can be compared with a value of 2006 K when 5 % argon is used as the thermal agent.

In order to change the location where heat extraction takes place, the calculation was repeated for a series of increasing E_a. Figure 56 is similar to Figure 55, but is for the case with E_a increased to 58.6 kJ/mol. The reaction behavior of X is very different than observed for the lower E_a. The concentration of X does not decrease appreciably until the temperature begins to increase at the boundary layer. As the temperature begins to rise on the oxidizer side, conversion of X to Y begins and accelerates with increasing temperature. A substantial fraction of the conversion takes place for temperatures greater than 1600 K, but, even so, the amount of unreacted X reaching the rich side of the flame is minimal. A maximum temperature of 1934 K is calculated. This is very close to the value found for $E_a = 25.1$ kJ/mol.

Figure 57 shows the results when the E_a is increased to 83.7 kJ/mol. For this large energy of activation, the conversion of X to Y has become so slow that a substantial fraction of X passes through the high temperature zone without conversion to Y. As the temperature falls on the rich side of the flame, the reaction slows down, and the remaining X becomes kinetically "frozen" and simply diffuses toward the fuel side. As a result, the amount of heat absorption is less than that occurring when X fully reacts. This is reflected in the maximum calculated flame temperature of 1963 K, which is intermediate between that found for X with lower E_a's and that for a 5 % argon volume fraction added to air, which only has dilution and heat removal effects.

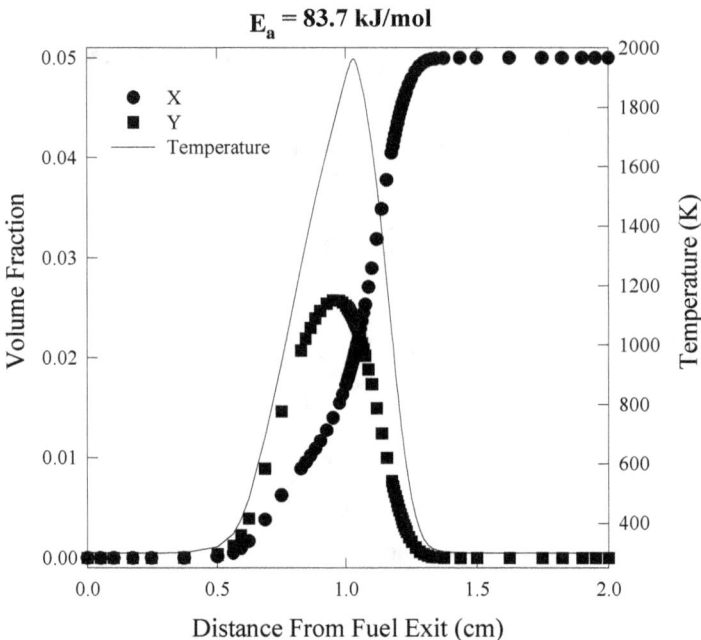

Figure 57. The volume fractions of X and Y are plotted as a function of distance from the fuel exit for a methane flame burning in air with 5 % added X. X reacts to form Y with E_a = 83.7 kJ/mol and $\Delta H_{X \rightarrow Y}$ = 96.1 kJ/mol. The spatial variation of temperature is also shown as the solid line. Methane and oxidizer exit velocity magnitudes are 25 cm/s.

Figure 58 shows the maximum calculated flame temperature as a function of assumed E_a for a series of calculations carried out for 5 % X added to the air. For $E_a \leq$ 50 kJ/mol the calculated temperature is constant within the variations expected for the different grids used for the calculations. As the E_a is increased further, the maximum temperature begins to increase. As we have seen, this is due to some fraction of X passing through the flame to the rich side without reacting to form Y. Since the temperature range over which the heat is absorbed by the reaction of X to Y varies widely as the E_a increases from 25 kJ/mol to 50 kJ/mol, these findings show that the maximum flame temperature and, by extension, flame extinction only depend on the amount of heat extracted and not where it occurs relative to the flame zone. It is concluded that the original hypothesis concerning the role of heat absorption location by gaseous thermal agents on extinguishment is not valid. This conclusion should be contrasted with the recent results of Lentati and Chelliah, who report that the location where heat extraction occurs can be important for flame inhibition when evaporating water droplets are added to the oxidizer flow. [73,74] Their findings seem to be associated with the fact that larger water droplets do not follow the oxidizer flow streamlines, whereas gaseous agents do.

A complete set of calculations was carried out for species X having $\Delta H_{X \rightarrow Y}$ = 96.1 kJ/mol and E_a = 41.8 kJ/mol. The results are plotted in Figure 59 as maximum calculated flame temperature versus exit velocity magnitudes for methane and oxidizer. Based on an extinction temperature of 1550 K, the extinguishing volume fraction for X in air is estimated to be 15.9 %. The corresponding value for argon was 43 %. Thus the heat extracted by the reaction of X to Y has reduced the amount of agent required by nearly 2/3. Figure 60 compares calculated values of maximum flame temperature as a function of velocity for 15 %, 30 %, and 45 % argon volume fraction with the corresponding results for 5 %, 10 %, and 15 % X. The two set of curves fall close together, but the agreement is not complete. At the lower concentrations the results for X fall slightly below those for Ar, while the opposite is true for the highest

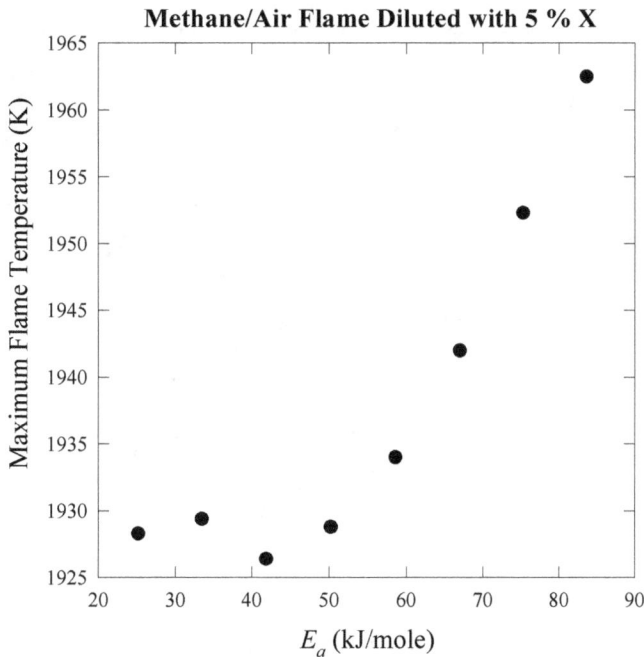

Figure 58. Values of maximum flame temperature for methane flames burning in air diluted with 5 % X are plotted as a function of the energy of activation for the conversion of X to Y. Methane and oxidizer exit velocity magnitudes are 25 cm/s.

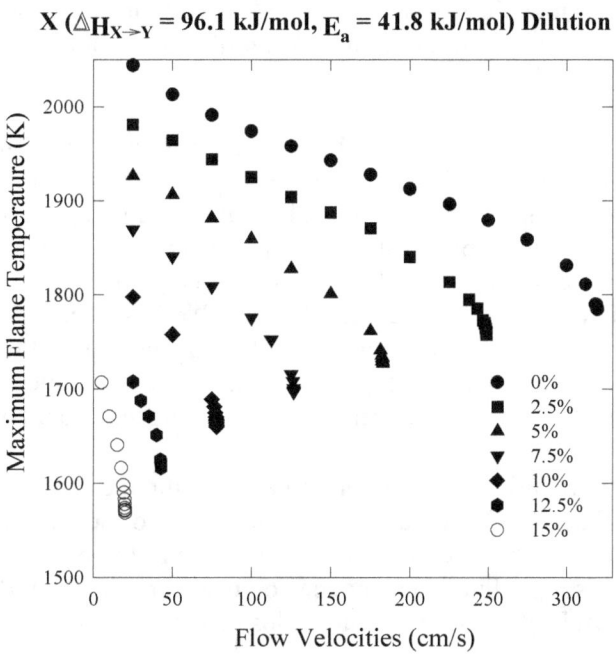

Figure 59. Maximum calculated flame temperatures are plotted against methane and oxidizer exit velocity magnitudes for a series of methane/air opposed-flow diffusion flames diluted with agent X that reacts to form Y with E_a = 41.8 kJ/mol and $\Delta H_{X \rightarrow Y}$ = 96.1 kJ/mol. Symbols correspond to the volume fractions of added X.

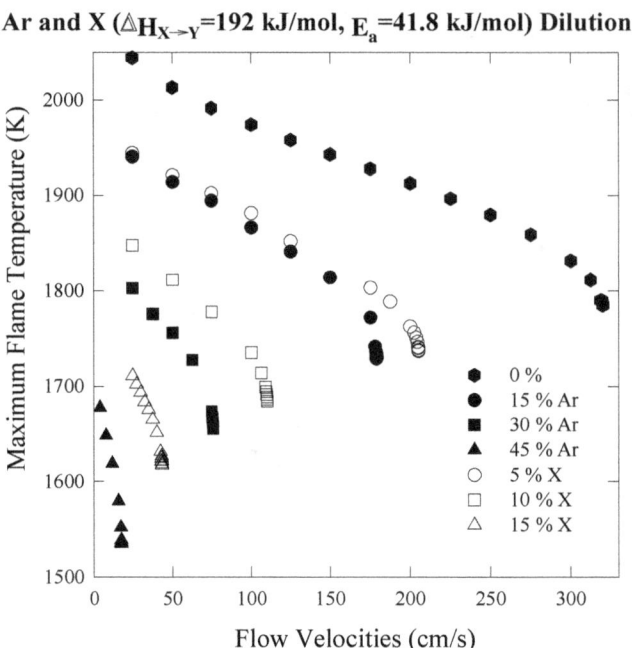

Figure 60. Maximum calculated flame temperatures are plotted against methane and oxidizer exit flow velocity magnitudes for a series of methane/air opposed-flow diffusion flames diluted with argon and agent X (E_a = 41.8 kJ/mol and $\Delta H_{X \to Y}$ = 96.1 kJ/mol).

concentrations. This behavior is most likely associated with the nonlinear dependence of the maximum flame temperature on concentration evident in Figure 44 and the contribution of dilution to extinguishment.

The effect of doubling the amount of heat absorbed by X was considered by running a series of calculations for $\Delta H_{X \to Y}$ = 192.1 kJ/mol. Calculated maximum flame temperatures are plotted as a function of the exit velocity magnitudes for methane and oxidizer in Figure 61. From these results the extinction volume fraction of X with the higher heat absorption can be estimated as 9.7 %. This value is roughly 60 % of that found with $\Delta H_{X \to Y}$ = 95.2 kJ/mol, or 20 % higher than would be expected if flame extinguishment was due solely to the heat extracted. The most likely source for the difference is the effect of dilution discussed earlier. The concentration of added agent required for flame extinguishment decreases as the amount of heat it can absorb increases, but the decreased concentration reduces the degree of dilution, with the result that the effectiveness of extracting additional heat is offset somewhat.

The role of dilution has an important implication with regard to simple estimates of extinguishing efficiency for thermal agents that are often obtained by taking ratios of heat capacities for various agents. If the agents have a large heat capacity difference, and the volume fractions required for extinguishment therefore differ substantially, a simple linear dependence on heat capacity should not be observed. In fact, the agent having the largest heat capacity should be less effective than expected, as observed in the current calculations.

Figure 62 shows a plot of maximum flame temperature at extinction versus exit flow velocity magnitudes for air diluted with various concentrations of argon and X with $\Delta H_{X \to Y}$ = 96.1 kJ/mol and 192.1 kJ/mol. The solid line is the result shown in Figure 43 based on a fit to data for known thermal agents. Despite the fact that the amount of agent required for extinction varies nonlinearly with $\Delta H_{X \to Y}$, the data all fall very close together. This supports the conclusion that assuming a common extinction temperature should

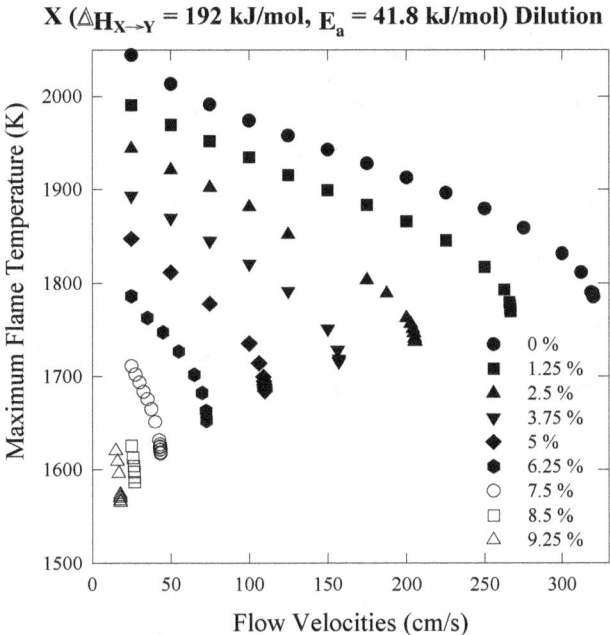

Figure 61. Maximum calculated flame temperatures are plotted against methane and oxidizer exit flow velocity magnitudes for a series of methane/air opposed-flow diffusion flames diluted with agent X which reacts to form Y with E_a = 41.8 kJ/mol and $\Delta H_{X \rightarrow Y}$ = 192.1 kJ/mol. Symbols correspond to the volume fractions of added X.

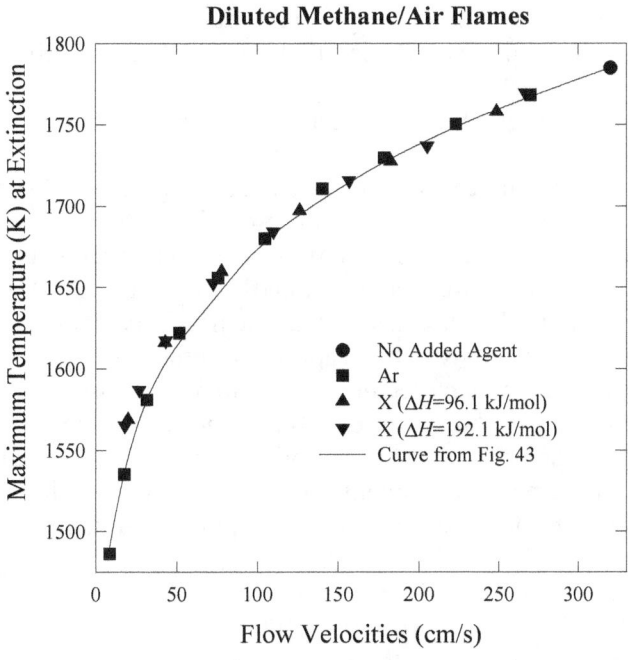

Figure 62. Values of maximum flame temperature at extinction are plotted as a function of fuel and oxidizer exit velocity magnitudes for air diluted with a range of volume fractions of argon and surrogate agent X having $\Delta H_{X \rightarrow Y}$ = 96.1 kJ/mol or 192.1 kJ/mol. The solid line is the result of the regression fits shown in Figure 43.

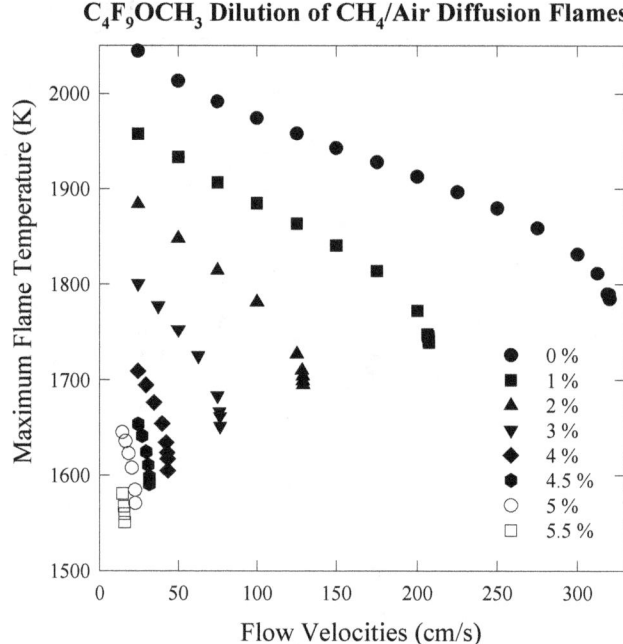

Figure 63. Maximum calculated flame temperatures are plotted against methane and oxidizer exit velocity magnitudes for a series of methane/air opposed-flow diffusion flames diluted with $C_4F_9OCH_3$. Symbols correspond to the volume fractions of added $C_4F_9OCH_3$.

provide excellent estimates for flame extinguishing volume fractions of thermal agents. The slight variations present near the extinguishing volume fractions are likely due to dilution effects.

I. Prediction of Extinguishing Concentration for HFE7100

The first section of this report summarizes data base searches that yielded a list of thermal agents that are expected to be particularly effective. In order to obtain some idea of the concentrations that might be required for extinguishment by such agents, a series of calculations were carried out for a compound that is predicted to be one of the more effective, methoxy-nonafluorobutane ($C_4F_9OCH_3$; HFE7100). It was assumed that the only mechanism for heat extraction was the heat capacity of the gaseous agent, even though significant heat could also be removed by evaporation if the agent were released as a liquid at room temperature. It is also true that this molecule is likely to decompose as it enters the high temperature region of the flame and will react to form final products. Reaction of the agent was not modeled. Due to the presence of a large number of fluorine atoms, it is also likely that the extinguishing efficiency will include a small chemical component. [65,67,68,85] On this basis it might be expected that the result of this calculation will be an upper limit for the required extinguishing concentration.

Figure 63 shows the maximum flame temperature as a function of the methane and oxidizer exit velocity magnitudes. As now expected, the curves have similar appearances to those found for the other thermal agents. However, what is particularly noteworthy is the relatively small volume fraction of $C_4F_9OCH_3$ required to lower the maximum flame temperature to the extinguishment temperature of 1550 K. The addition of only 5.5 % of $C_4F_9OCH_3$ is calculated to be sufficient. The high heat capacity of this molecule due to its large number of atoms makes it a particularly effective thermal agent in molar terms. If liquid vaporization and chemical effects improve its effectiveness still further, this species may well approach the effectiveness of halon 1301, which has an extinguishing concentration of 2.9 % [54] on a molar basis. On a mass basis this fluorinated ether will be less effective, based on the molecular weights of 148.9 g/mol for halon 1301 and 250 g/mol for $C_4F_9OCH_3$, but not dramatically so. It might be expected that the mass fraction of $C_4F_9OCH_3$ that must be added would be similar to other agents, which do not have a strong component of chemical effectiveness. [33]

Using the same assumptions as for the diffusion flame calculations, the adiabatic flame temperature for a premixed methane flame burning in air diluted by an extinguishing concentration of $C_4F_9OCH_3$ was calculated to be 1696 K. This is considerably lower than the corresponding temperatures calculated for extinguishing concentrations of N_2, CO_2, Ar, and He (see Table 18).

The vapor pressure for $C_4F_9OCH_3$ as a function of temperature is given by [23]

$$P_{C_4F_9OCH_3} = e^{10.839 - \frac{3627.2}{T}}, \quad (20)$$

where $P_{C_4F_9OCH_3}$ is the partial pressure of $C_4F_9OCH_3$ in atmospheres and T is the temperature in K. For 300 K and atmospheric pressure the saturation volume fractions of the ether in air is calculated to be 29 %. Thus the saturation pressure at room temperature should be more than sufficient to extinguish a diffusion flame.

Two experimental measurements of the volume fractions of methoxy-nonafluorobutane required to extinguish diffusion flames have been identified. Unpublished measurements from the New Mexico Engineering Institute using a standard cup burner with heptane fuel yielded an extinguishing volume fraction of 6.1 %. [140] In a patent disclosure, Flynn and Scott also reported that a volume fraction of 6.1 % was sufficient to extinguish a butane flame in a "micro-cup burner". [141] These values are 11 % higher than estimated using detailed chemical-kinetic modeling for the extinguishment of a methane flame.

Earlier it was shown that cup burner measurements using heptane fuel for agents known to act only thermally yielded extinguishing concentrations that were roughly 90 % to 95 % of those estimated using detailed kinetic modeling. This would suggest that the expected difference between calculated and experimental values for $C_4F_9OCH_3$ are actually somewhat greater than indicated above. A plausible explanation for this difference is the effect of chemical reactions. If this is the indeed the case, the results indicate that the net effect of reactions involving this agent is to decrease its effectiveness slightly from that expected if it acted simply as a nonreactive thermal agent. Other reaction effects must counteract the expected small chemical enhancement due to the fluorine atoms.

IV. SURFACE COOLING

A. Introduction

When liquid suppressant droplets released from a discharge port approach a fire, several things can happen. Droplets with sufficient momentum to penetrate the flame, depending on their sizes, can either be consumed in the flame (participate directly in the suppression processes through physical and/or chemical mechanisms) or reach the fuel surface. Droplets with little momentum (*e.g.*, very small droplets) may not penetrate the fire and will be deflected away by the rising hot plume. [142] These deflected droplets will eventually evaporate or strike adjacent objects. For droplets landing outside the burning area, the cooling of the adjacent and surrounding surfaces could mitigate or contain the flame spread processes. Therefore, droplet interaction with surfaces is an integral part in the understanding of the overall suppression processes.

One of the possible fire suppression mechanisms using a liquid fire suppressant (*e.g.*, water) is fuel surface cooling. The relatively cool impacting droplets will interact with the relatively hot fuel surface. One of the droplet/surface interactions is the extraction of heat from the surface to vaporize the continuously deposited droplets, thereby reducing the surface temperature. The burning rate falls, which leads to a decrease in heat feedback to the fuel surface. Eventually, a critical burning rate is reached beyond which combustion can no longer be sustained. The other effect of the deposited droplets on the burning rate is the reduction of mass transfer area on the fuel surface, which also causes a reduction in burning rate. For solid fuels these processes are generally regarded as the dominant mechanism for extinguishment by water sprays. [143]

In order to understand surface cooling without resorting to a detailed analysis of the coupling between the transient thermal response of the fuel bed as a result of the application of liquid agent and the heat feedback from the flame to the fuel, the transient coupling process is considered as a series of quasi-steady steps. At any instant in time, a steady-state energy balance at the fuel surface can be written as:

$$\dot{m}''_{fuel} \Delta H_{fuel} = Q''_{cond} + Q''_{conv} + \varepsilon_{flame} \sigma T^4_{flame} - \varepsilon_s \sigma T^4_{surf} - q'' \qquad (21)$$

where $\dot{m}''_{fuel}$ is the fuel burning mass flux, ΔH_{fuel} is the effective heat of gasification of the fuel, Q''_{cond} is the conductive heat flux to the fuel surface, Q''_{conv} is the convective heat flux to the fuel surface, ε_{flame} is the flame emissivity, σ is the Stefan-Boltzmann constant, T_{flame} is the average flame temperature, ε_s is the surface emissivity, T_{surf} is the fuel surface temperature, and q'' is the heat flux removal from the fuel surface due to surface cooling.

The surface cooling heat flux, q'', depends on the interaction of the liquid droplets with the surface. There are many different scenarios of droplet impact on a surface. The impacting processes depend on the thermophyscial and geometrical properties of the drop and of the target surface and on the magnitude and direction of the impact velocity vector. In addition, for a flaming surface, the droplet will traverse through an environment different from that of a non-burning surface before impact. However, in the literature, droplet/surface interaction studies focus almost exclusively on non-burning surfaces.

When relatively cool liquid droplets impact on a solid surface, cooling of the surface occurs as a result of heat transfer between the surface and the droplets. A liquid droplet impacting a solid surface has been extensively studied in the literature. [144,145,146,147,148,149,150,151,152] Maps that identify the various impact and heat transfer regimes over a wide range of experimental conditions have been constructed. Spray cooling or quenching of hot metallic surfaces has also been extensively studied. [153,154,155,156,157,158] Many empirical heat transfer correlations for spray cooling applications exist.

For a liquid droplet interaction with a *liquid* surface, several scenarios need to be considered, and the impact dynamics are complicated by the ability of the liquid surface to deform and to displace in response to the impinging droplet and by the internal fluid motion of the liquid pool initiated by the impact processes. [159] If the droplet has a very high impact velocity, it can cause a crater in the liquid pool and a splash upon impact. For a burning liquid pool, the splash may result in a sudden increase in burning rate due to an increase in the liquid surface area caused by the splash droplets. If the droplet is soluble in the liquid, mass transfer between the penetrating droplet and the liquid will occur. Depending on droplet impact parameters, differences in miscibilities and physical properties (surface tension and viscosity) between the droplet and the target liquid, and the depth of the liquid layer, the droplet may float on the surface, may coalesce into the liquid pool with little or no splashing and produce a vortex ring, may rebound or splash, forming a crater and a small jet or column of liquid at the surface, or may dissolve into the liquid pool. Most work in the literature has examined the isothermal impact of a liquid droplet onto a surface of the same liquid (e.g., see the review article of Rein [159] and the references therein). In a fire suppression scenario, the impacting droplet encounters a different liquid target (suppression fluid versus fuel) with different physical properties. The temperature of the droplet can be much lower than that of the liquid surface (especially, in the case of burning), and the contact between a cold droplet with a hot liquid may sometimes results in vapor explosion [160], an unwanted situation. All of these complicated factors render the study of heat transfer processes between liquid droplets and a liquid surface an arduous and daunting task. To the best of our knowledge, appropriate heat transfer correlations do not exist in the literature; therefore, liquid surface cooling calculations, in spite of the importance of the process, will not be attempted here.

In this section, we will discuss droplet evaporation and solid surface cooling by droplets. Droplet evaporation is considered because it is an integral part of the surface cooling problem. If the droplets evaporate completely before reaching the fuel surface, the heat removal from the surface will decrease significantly due to the fact that the liquid latent heat of vaporization does not play a role. Under this circumstance, the surface will only be cooled by a vapor jet, which is not an effective cooling mechanism.

Another important aspect in surface cooling is the ability of the droplets to reduce the surface temperature of metallic components in the vicinity of a fire to prevent re-ignition, which is a potential problem in suppressing aircraft engine nacelle fires. If sufficient cooling of hot surfaces cannot be achieved after the release of fire suppressant, re-ignition of residual fuel vapor/air mixture by hot surfaces may occur.

In the following discussion, the five fluids (water, lactic acid, $C_3F_5H_3O$, HFE7100, and R338mccq) identified in Section II.A as having the highest ΔH^{total} (mass basis) are considered to evaluate their droplet vaporization and surface cooling characteristics. Low boiling-point fluids will not be considered here because under typical operating conditions, these fluids can be easily flash-vaporized soon after they leave the discharge opening; thus they likely arrive at the surface in the form of a vapor.

B. Evaporation of Liquid Droplets

The estimation of liquid droplet evaporation is based on the classical d^2-law, which gives the droplet diameter as a function of time as

$$D^2 = D_o^2 - Kt \quad (22)$$

where D is the instantaneous droplet diameter, D_o is the initial droplet diameter, t is the time, and K is the evaporation constant. The evaporation constant can be calculated by

$$K = \frac{8k_g \ln(1+B)}{\rho_F C_{pg}}, \quad (23)$$

where B is the transfer number, k_g is the gas-phase thermal conductivity, C_{pg} is the gas-phase heat capacity, and ρ_F is the liquid density. The B number based on mass transfer considerations is given by

$$B_M = \frac{Y_{FS} - Y_{F\infty}}{1 - Y_{FS}}, \quad (24)$$

where Y_{FS} and $Y_{F\infty}$ are the vapor-phase mass fraction at the droplet surface and ambiance, respectively. The mass fraction, Y_{FS}, can be evaluated by

$$Y_{FS} = \frac{P_{FS} M_F}{P_{FS} M_F + (P - P_{FS}) M_A}, \quad (25)$$

where P_{FS} is the liquid vapor pressure at the droplet surface, P is the ambient pressure, which is equal to the sum of the liquid vapor pressure and partial pressure of air at the droplet surface, and M_F and M_A are the molecular weights of liquid and air, respectively.

The B number based on heat transfer considerations is given by

$$B_T = \frac{C_{pg}(T_\infty - T_s)}{\Delta H_{vap}}, \quad (26)$$

where T_∞, T_s, and ΔH_{vap} are the ambient temperature, droplet surface (wet-bulb) temperature, and latent heat of vaporization of the liquid, respectively. Under steady-state conditions, $B_M = B_T$, and either Eq. (24) or Eq. (26) may be used to calculate the evaporation constant in Eq. (23).

If T_s is known, the calculation of the evaporation constant is straightforward; however, under normal circumstances it is necessary to determine T_s in order to evaluate B and K. The calculation procedure can be found in references [161,162,163]. The basic idea is to find T_s such that $B_M = B_T$. Spalding [161] and Kanury [162] used a graphical approach. By plotting B_M and B_T against T_s, the intersection of these two curves defines the wet-bulb temperature and the corresponding B. An iterative method was used by Chin and Lefebvre. [163] The calculations presented here use the latter approach.

To calculate K, the reference temperature (T_r) and mass fractions (Y_{Fr} and Y_{Air} for the fluid and air, respectively), based on the one-third rule [163], were used to estimate the gas-phase thermophysical properties,

$$T_r = T_s + \frac{T_\infty - T_s}{3} \quad (27)$$

$$Y_{Fr} = Y_{FS} + \frac{Y_{F\infty} - Y_{FS}}{3} \quad (28)$$

$$Y_{Air} = 1 - Y_{Fr}. \quad (29)$$

The equation for calculating the gas-phase heat capacity, C_{pg}, is given by

$$C_{pg}(T_r) = Y_{Air} C_{pA}(T_r) + Y_{Fr} C_{pF}(T_r) \quad (30)$$

where C_{pA} is the heat capacity of air and C_{pF} is the gas-phase heat capacity of the fluid, all evaluated at T_r. The thermal conductivity of the gas phase is estimated by the following equation,

$$k_g(T_r) = Y_{Air} k_A(T_r) + Y_{Fr} k_F(T_r), \quad (31)$$

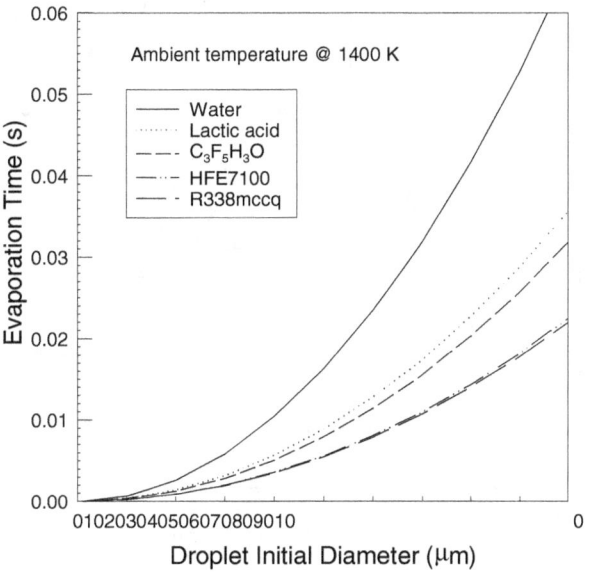

Figure 64. The droplet evaporation time is shown as a function of initial droplet diameter for the five fluids indicated.

where k_A is the thermal conductivity of air and k_F is the gas-phase thermal conductivity of the liquid, all evaluated at T_r.

The conditions used in the calculations were: (1) $T_\infty = 1400$ K (to simulate a flaming ambiance), (2) $Y_{F\infty} = 0$, and (3) $P = 0.101$ MPa.

The thermophysical properties of air, water, and lactic acid were obtained from the database of the Design Institute for Physical Properties. [9] The properties of $C_3F_5H_3O$, HFE7100, and R338mccq were estimated by methods described in Reid *et al.* [164]; the modified Rackett technique was used for ρ_F, the method of Joback for C_{pF}, the method of Lee and Kesler for P_{FS}, the method of Pitzer's acentric factor correlation for ΔH_{vap}, and the method of Ely and Hanley for k_F.

Figure 64 shows droplet evaporation times, t_b, as a function of initial droplet diameter. The droplet evaporation time can be easily obtained by assigning $D = 0$ in Eq. (22) once K is known,

$$t_b = \frac{D_o^2}{K}. \qquad (32)$$

For a given initial droplet diameter under the same ambient conditions, the evaporation time for a water droplet is the longest and for a R338mccq droplet the shortest. For the five fluids examined, the ranking based on droplet evaporation times is

$$t_{b,R338mccq} < t_{b,HFE7100} < t_{b,C_3F_5H_3O} < t_{b,lactic\,acid} < t_{b,water}. \qquad (33)$$

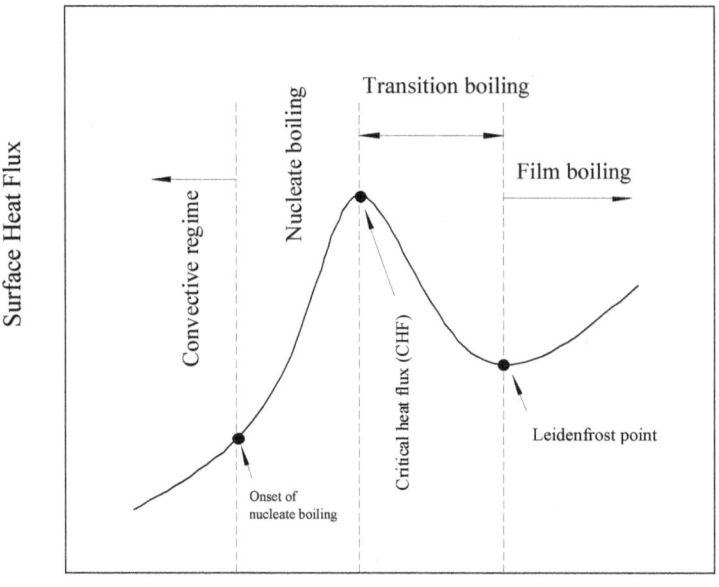

Figure 65. An illustration of a typical boiling curve associated with quenching of a hot surface by liquid droplets is shown.

How fast a droplet evaporates determines whether it will still remain as a droplet upon reaching the surface. However, if *flame cooling* is the dominant fire suppression mechanism, then it would be better to have the droplet completely vaporized in the flame.

C. Spray Cooling

In order to assess surface cooling, we consider the literature on spray cooling and quenching of hot metallic surfaces. The application of spray-cooling correlations to fuel surface cooling is only appropriate when the fuel surface is solid (*e.g.*, a PMMA or wood slab). Even in this case, care should be exercised because subtle differences in cooling mechanisms exist. A burning PMMA surface in reality is in a molten state, and wood forms a porous char layer. The interactions of droplets with molten and porous surfaces remain very difficult problems to tackle. Differences also arise between fuel surface cooling and quenching of metallic surfaces. In the latter, radiative heat transfer to the droplets only occurs from the heated surface; however, in the former, it also occurs from the flame. Strictly speaking, the following calculations are more applicable to the cooling of a hot surface to prevent re-ignition than to the cooling of a burning fuel surface.

Depending on the surface temperature, when a hot surface is being quenched by a liquid spray, it will experience several distinct heat transfer regimes which can be followed along the "boiling curve". Such a curve is a plot of surface heat flux versus surface temperature and represents the strong relationship between heat flux and surface temperature during quenching. The characteristics of the boiling curve are classified as (with increasing surface temperature): (1) convective, (2) nucleate boiling, (3) critical heat flux, (4) transition boiling, (5) Leidenfrost point, and (6) film boiling. Detailed descriptions are available. [153,154] Figure 65 is an illustration of a boiling curve associated with spray cooling of a hot surface. The general features of the curve are similar to those for a pool boiling curve.

Since the boiling curve of one fluid may differ from that of other fluids, there are several important assumptions that have to be made in the following analysis.

When a comparison of surface heat fluxes is made, it is assumed that for the prevailing surface temperature, these fluids are in the *same* regime on the boiling curve. This assumption may not be valid because, for example, the surface temperatures at which the critical heat flux and Leidenfrost phenomena occur may be significantly different among the fluids studied; therefore, while one fluid may be at the critical heat flux, the other fluid may be in the nucleate boiling regime at the prevailing surface temperature. Unfortunately, in most cases, it is not possible *a priori* to determine the demarcation point (surface temperature) from one cooling regime to another without resorting to experimental observations.

The second assumption is that the dimensionless heat transfer correlations obtained in the literature using water (in most cases) are applicable to the other fluids considered here. A third assumption is that the effect of surface roughness on surface heat flux is not considered. Furthermore, the spray parameters used in the following calculations were chosen to be within the range of applicability of the empirical correlations. These operating parameters, though encountered in spray quenching applications, may not be identical to those appropriate for fire fighting applications. In addition, since different droplet generation techniques (sprays *vs.* mono-dispersed droplet streams) have been used to obtain the heat transfer correlations in various regimes, it is imperative that a comparison among the calculated surface heat fluxes should be made with these fluids in the *same* regime.

D. Convective Regime

Mudawar and co-workers [153,154] have performed extensive studies on cooling of metal surface in this regime using water sprays. Their heat transfer data were correlated using the Nusselt number (Nu), the Reynolds number (Re), and the Prandtl number (Pr) by the following expression,

$$\frac{hD_{32}}{k_f} = Nu = 2.512 \left[\frac{Q''D_{32}}{v_f}\right]^{0.76} \left[\frac{C_{pf}\mu_f}{k_f}\right]^{0.56} = 2.512 Re^{0.76} Pr^{0.56}, \quad (34)$$

where h is the heat transfer coefficient, k_f is the thermal conductivity of the liquid, v_f is the liquid kinematic viscosity, C_{pf} is the liquid heat capacity, and μ_f is the liquid viscosity. The Reynolds number is based on the volumetric spray flux, Q'', and the Sauter mean diameter, D_{32}, of the spray. All the liquid properties are evaluated at $(T_{surf} + T_f)/2$ where T_f is the droplet temperature before impact.

Figure 66 shows the calculated heat flux as a function of $T_{surf} - T_f$. The calculations were performed using $Q'' = 0.005$ m^3/m^2-s, $T_f = 295$ K, and $D_{32} = 0.5$ mm. The thermophysical properties of water and lactic acid were obtained from the DIPPR. [9] The properties of C$_3$F$_5$H$_3$O, HFE7100, and R338mccq were estimated by methods described in Reid *et al.* [164]; the boiling point method of Sato was used for k_f, the corresponding states method of Rowlinson for C_{pf}, and the method of Brule and Starling for μ_f. In this regime, water removes heat from the surface more efficiently than the other four fluids. It should be noted that it may not be feasible in practice to maintain the same D_{32} with a fixed Q'' (the conditions used in the calculations) for the five fluids because the atomization characteristics at the spray nozzle would be vastly different among the fluids owing to their thermophysical properties, thus resulting in different D_{32}. [165]

E. Nucleate Boiling Regime

Compared to other regimes on the boiling curve, the nucleate boiling regime in spray cooling is probably the least studied. In this regime, the heat flux removal from the surface was found to scale only with $T_{surf} - T_f$ in the studies by Mudawar and co-workers [153,154] where a water spray was used. The heat flux, q'', was correlated in the form of

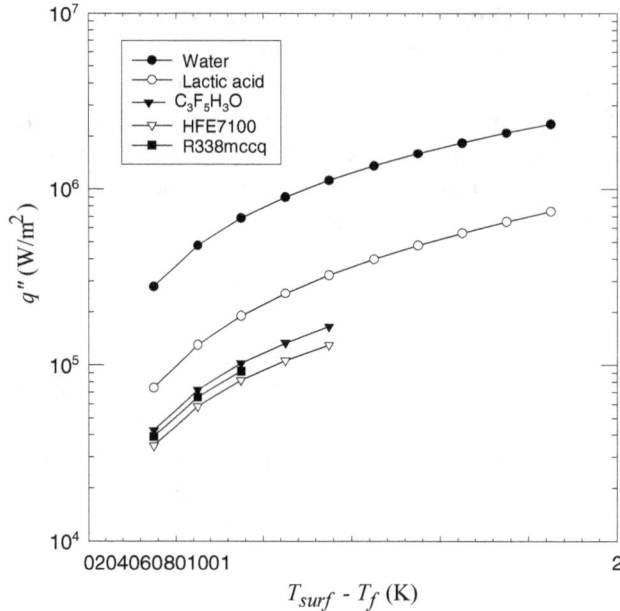

Figure 66. Calculated boiling curves in the convective regime are shown for the five indicated liquids.

$$q''=1.87\times 10^{-5}(T_{surf}-T_f)^{5.55}. \tag{35}$$

In the work of Qiao and Chandra [166] on nucleate boiling enhancement using a surfactant/water solution, the heat flux measurements could also be correlated well by scaling with $T_{surf} - T_f$ in a form similar to Eq. (35),

$$q''=0.56\times 10^{-5}(T_{surf}-T_f)^{6}. \tag{36}$$

If we *assume* that the proportionality constant and the power in the scaling law do not change significantly among the five chemical compounds and that the heat flux can be scaled with (T_{surf} - T_f), then the heat flux removal from the surface in the nucleate boiling regime using these five liquids should be similar for a given T_{surf}.

F. Critical Heat Flux

The dimensionless correlation developed by Mudawar and Valentine [153] for water spray cooling over a wide range of flow rates is used to estimate the critical heat flux (*CHF*),

$$\frac{CHF}{\rho_g \Delta H_{vap} Q''}=122.4\left\{1+0.0118\left[\frac{\rho_g}{\rho_F}\right]^{1/4}\left[\frac{\rho_F C_{pf}(T_{sat}-T_f)}{\rho_g \Delta H_{vap}}\right]\right\}\left(\frac{\sigma_f}{\rho_F Q''^2 D_{32}}\right)^{0.198}, \tag{37}$$

where ρ_g is the vapor density of the fluid, T_{sat} is the saturation temperature of the fluid, and σ_f is the surface tension of the fluid. All the fluid properties are evaluated at the fluid saturation temperature.

Using $Q'' = 0.005$ m^3/m^2-s, $T_f = 295$ K, and $D_{32} = 0.5$ mm, the resulting *CHF*s for water, lactic acid, C$_3$F$_5$H$_3$O, HFE7100, and R338mccq are 0.63×10^7 W/m^2, 0.86×10^7 W/m^2, 0.33×10^7 W/m^2, 0.23×10^7

W/m², and 0.21 × 10⁷ W/m², respectively. In the calculations, surface tensions of water and lactic acid were obtained from the DIPPR [9]; while those of C$_3$F$_5$H$_3$O, HFE7100, and R338mccq were estimated using the method of Brock and Bird. [164] Although lactic acid and water have the highest and second highest *CHF*s, respectively, under the conditions used in the calculations, it should be noted that it is inappropriate to compare the *CHF*s unless they occur at the same surface temperature for the five fluids, a situation highly unlikely given the differences in the thermophysical properties of these fluids.

G. Transition and Film Boiling Regimes

Based on dimensional analysis, Deb and Yao [167] statistically derived an equation for heat transfer effectiveness (ε) in terms of droplet Weber number (*We*), dimensionless wall superheat (B_w) and vapor parameters (K_d), and surface factor (*SF*) for surface material effect,

$$\varepsilon = 0.027 \exp\left[\frac{0.08\sqrt{\ln\left(\frac{We}{35}+1\right)}}{\left(B_w + \frac{SF}{60.5}\right)^{1.5}}\right] + 0.21 K_d B_w \exp\left[-\frac{90}{We+1}\right], \quad (38)$$

with $\quad We = \dfrac{\rho_F V_i D_o}{\sigma_f}, \quad B_w = \dfrac{C_{pF}(T_{surf} - T_{sat})}{\Delta H_{vap}}, \quad K_d = \dfrac{k_F}{C_{pF} \mu_v}, \quad$ and

$$SF = \sqrt{\frac{k_s \rho_s C_{ps}}{(k\rho C_p)_{steel}}} - 1,$$

where V_i is the droplet impact velocity, K_d is the dimensionless vapor parameter, μ_v is vapor viscosity, and k_s, ρ_s, and C_{ps} are the thermal conductivity, density, and heat capacity of the surface material, respectively. All liquid and vapor properties are evaluated at T_{sat}. The heat transfer effectiveness is defined as the ratio of actual heat transfer from the hot surface to the total heat transfer required for complete evaporation of the droplets.; therefore, the heat flux from the surface is

$$q'' = \varepsilon G[\Delta H_{vap} + C_{pf}(T_{sat} - T_f)], \quad (39)$$

where G is the liquid mass flux. ε was obtained from Eq. (38) by fitting data mainly from experiments using a single stream of droplets. The experimental conditions covered the film boiling regime and to some extent the adjacent transition boiling regime and Leidenfrost point.

Figure 67 shows the calculated heat fluxes for the five fluids as a function of the surface temperature. The calculations were performed using T_f = 295 K, V_i = 2.5 m/s, D_o = 0.3 mm, and G = 0.5 kg/m²-s. For illustration, a stainless steel, SS 304, was used as the target surface. Under the conditions used in the calculations, the resulting droplet *We* are 31, 97, 181, 297, and 339 for water, lactic acid, C$_3$F$_5$H$_3$O, HFE7100, and R338mccq, respectively. The range of surface temperatures used in the calculations was chosen so that the calculated ε did not exceed 1. The three boiling regimes (transition boiling, Leidenfrost condition, and film boiling) are apparent in Figure 67. If we assume that at $T_{surf} \geq 800$ K, all five fluids exhibit film boiling, and lactic acid has the highest surface heat flux.

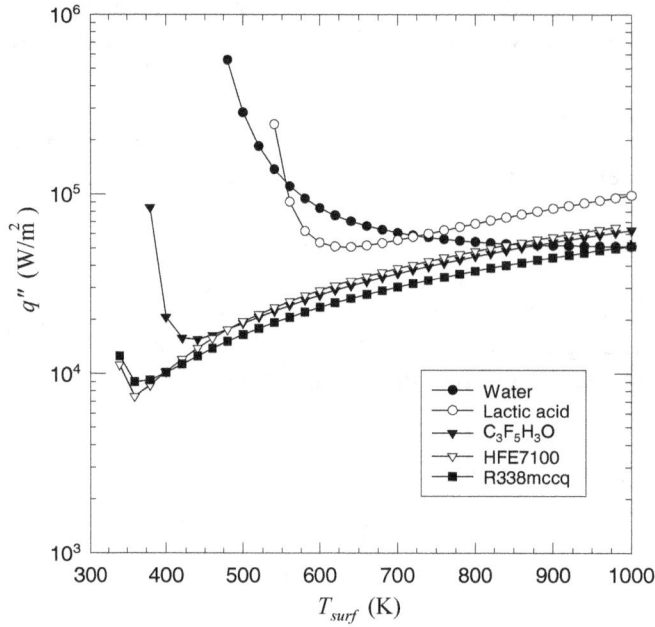

Figure 67. Calculated boiling curves for the five liquids are shown for the transition and film boiling regimes.

Figure 67 also demonstrates again the difficulty in comparing the heat fluxes at a given surface temperature. For example, at T_{surf} = 550 K, $C_3F_5H_3O$, HFE7100, and R338mccq are in the film boiling regime, whereas, water and lactic acid are in the transition boiling regime. As stated above, unless the surface temperature at which the onset of a particular boiling regime is known *a priori*, it is not possible to know which heat flux correlation pertaining to which boiling regime should be applied in the estimation.

H. Summary

Empirical heat transfer correlations from the literature have been used to assess surface cooling in various heat transfer regimes. Based on the above calculations, *water* and *lactic acid* generally appear to be superior to $C_3F_5H_3O$, HFE7100, and R338mccq in *surface cooling* applications. Water and lactic acid evaporate much slower than $C_3F_5H_3O$, HFE7100, and R338mccq; therefore, the likelihood for water and lactic acid droplets to reach the surface is higher. For the convective boiling regime, the calculations show that water is better than lactic acid, $C_3F_5H_3O$, HFE7100, and R338mccq. Lactic acid has the highest calculated critical heat flux as well as the highest heat flux in the film boiling regime. In order to validate the calculated ranking of surface cooling effectiveness of these agents, experiments should be performed in various heat transfer regimes under conditions commensurate with those in actual applications.

Surface cooling by liquid droplets is a very complicated heat transfer process because it depends on so many parameters which include the thermophysical properties (surface tension, density, viscosity, heat of vaporization, etc.) of the liquid, droplet size, droplet impact velocity, spray mass flux, and surface properties. Even for non-burning, smooth surfaces, calculations of surface heat fluxes from first principles are not possible without resorting to experimental observations. Empirical heat transfer correlations under conditions encountered in fires are lacking, and the quantification of surface cooling as a means to extinguish the burning fuel remains elusive, especially when the fuel is a liquid.

V. DISCUSSION

A. Liquid HFE7100 as a Potential Fire-Fighting Agent and Liquid Agent Effectiveness

One of the major goals of this project was to identify highly effective thermal agents that might serve as replacements for halon 1301. Based on a thermodynamic analysis and practical considerations such as physical properties and commercial availability, HFE7100 was identified as being particularly promising. As a purely thermal agent it ranked near the top of the list of potential replacement agents on both molar and mass bases (see Table 5 and Table 6). Measurements in the Distributed Liquid Agent Fire Suppression Screen apparatus yielded effective extinguishing concentrations of 1.5 % and 11.7 % in volume fraction and mass fraction terms, respectively, for a propane flame burning on a Tsuji-type burner. Note that the HFE7100 was released as a liquid, while the extinguishing concentrations are those obtained by assuming the agent is totally vaporized.

The results of the detailed chemical-kinetic modeling study suggested that the effectiveness of gaseous thermal agents does not depend on the location of heat extraction relative to the flame front as long as the heat extraction is complete and the gases are convected into the flame zone. If the same holds true for liquid agents, it should be possible to estimate the required effective liquid concentration assuming full vaporization of the agent and using a thermal analysis similar to that suggested by Eq. (1). Based on the data included in Table 2, Table 4, Table 5, and Table 6 and plotted in Figure 1 and Figure 2, it is apparent that the largest fraction of heat absorption by HFE7100 takes place due to gas-phase heating, with liquid-phase processes contributing roughly 10 % of the overall heat extraction. The detailed chemical-kinetic modeling study of the extinction behavior of a methane counterflow flame by gas-phase HFE7100 yielded an estimated extinguishing volume fraction of 5.9 %, which was slightly less than suggested by unpublished experimental measurements. The modeling study also indicated that roughly 20 % more thermal agent was required to extinguish a propane flame as for a methane flame. On this basis, we estimate that the extinguishing volume fraction of gaseous HFE7100 required for a propane flame should be on the order of 7.1 %. Reducing this by 10 % to account for the heat absorbed by the liquid yields an estimate of 6.4 % for the effective volume fraction of liquid agent required to extinguish a propane flame. This value is more than four times larger than observed experimentally.

Despite the wide spread use of fire-fighting agents released as liquids, note that liquid water remains the most frequently used means for fire fighting and that halon 1301 is generally released as a pressurized liquid, there are very few studies that have considered the direct interaction of a flame with liquid droplets such as reported here. This is particularly surprising given the importance of water for practical fire fighting. In a very early study, Seshadri investigated the extinguishment of heptane, methanol, and wood flames in counterflow configurations using water droplets. [168] The droplets were small enough that they were considered to have evaporated before reaching the flame surfaces. Ewing et al. cite unpublished values for water falling in the range of 5.5 % to 13.3 % for the effective volume fraction of water (assuming complete evaporation) for diffusion flames. [61] The only other studies of which we are aware have been carried out as part of the NGP and have investigated water. Similar to the current investigation, these studies were performed in counterflow flames using strain rates that were considerably higher than are characteristic of buoyancy-dominated flames. In order to estimate extinguishing concentrations it is necessary to extrapolate the findings to appropriate lower strain rates.

Zegers et al. have reported extinction results for a propane opposed-jet diffusion flame using various sized water droplets. [169]. The effectiveness of the water for inhibiting the flame varied with nominal droplet diameter. Simple extrapolation of the results to strain rates characteristic of buoyancy-dominated flames provides estimates of 1.3 %, 4.6 %, and 5.2 % in terms of mass fractions (2.1 %, 6.4 %, and 8.1 % in terms of volume fractions) for droplets having nominal diameters of 14 μm, 30 μm, and 42 μm, respectively. Similar measurements for a propane flame were made by Yang et al. using the DLAFSS. [35] Their result in mass fraction terms is 5.2 %, which corresponds to a volume fraction of 8.1 %, for nominally 30 µm-diameter water droplets. (Note that a factor of two has been incorporated into this result to reflect the coverage factor for the spray in the DLAFSS.) Lazzarini et al. have reported similar extinction measurements for a methane/air opposed-jet flame using droplets with a median diameter of 20 µm. [170] Extrapolation of their results leads to estimated extinguishing concentrations of 3.3 % (mass fraction) and 5.2 % (volume fraction). Even though these estimates must be considered very crude and do not consider fuel effects, there is general agreement between the three investigations.

From the detailed chemical-kinetic modeling the extinguishing volume fraction of water vapor in air for a methane flame has been estimated as 28 %. Using the data included in Table 2, Table 4, Table 5, and Table 6 and plotted in Figure 1 and Figure 2, it can be estimated that roughly one half of the heat extracted by heating water from room temperature to flame temperatures arises from heating of the liquid and vaporization. Based simply on thermal considerations, one would therefore predict nominal volume fractions of 14 % for extinguishment of methane flames and 1.2 times higher, or nearly 17 % for propane flames. Again, these simple arguments result in overestimates as compared to the extrapolated experimental values by factors that range from three to thirteen.

Experimental extinguishment measurements using liquid agents are difficult and are subject to substantial uncertainties. Furthermore, it is necessary to extrapolate from relatively high strain rates to the lower strain rates characteristic of buoyancy-dominated flames. However, based on the existing data, it does appear that liquid agents are considerably more effective than would be predicted based simply on their ability to extract heat from a flame zone. Other physical processes may be playing a role. If such effects are actually present, it may be possible to tailor such effects to improve the effectiveness of liquid agents. Additional research is required to confirm the results discussed here and to provide an understanding for the mechanisms that lead to the observed improvements in liquid agent effectiveness.

Even though liquid agents appear to be more effective than expected in extinguishing opposed-flow diffusion flames, the opposite was found to be the case for HFE7100 in the TARPF. Here a nominal volume fraction of 2.8 % HFE7100 was insufficient to extinguish the flame. This value is nearly a factor of two higher than required to extinguish the flame in the DLAFSS. The difference in the two experiments is almost certainly associated with agent distribution and entrainment in the TARPF. The volume fraction of HFE7100 cited is a nominal value, and it is unclear how evenly distributed the agent was in the wind tunnel. There may be regions in the cross section near the flame front that have actual HFE7100 concentrations below those required for flame extinguishment. It is also possible that efficient extinguishment requires entrainment of the agent into the recirculating zone located behind the bluff body used for flame attachment in the TARPF. It is likely that droplets of agent dispersed in the high-speed air flow have sufficient momentum to pass above the recirculating zone without being entrained. In this way, it would be possible to have an extinguishing concentration of liquid agent in the air above the flame, but a lower concentration in the actual location where the flame is stabilized. In either case, the results demonstrate the importance of agent distribution for effective flame extinguishment.

Based on the thermodynamics analysis and experimental results reported in the current work it is concluded that HFE7100 warrants further consideration as a potential replacement for halon 1301. It should be noted that other compounds included in Table 5 and Table 6 also have the potential to be effective thermal fire extinguishing agents. Even though the search for chemicals was extensive, it is also possible that similar compounds might have been overlooked. We focused on HFE7100 because it appeared to be highly promising on both a mass and a volume fractions basis, was available commercially, and had already received SNAP approval from the EPA for another use.

B. Conclusions from Detailed Chemical-Kinetic Modeling of Thermal Agent Effects

Even though flame extinguishment has been extensively studied, the short review in Section III.A demonstrates that models for flame extinguishment by thermal agents are not well developed and that contradictory approaches have been discussed. Detailed chemical-kinetic modeling of counterflow diffusion flames has been shown to be an effective approach for improving the understanding of the effects of thermal agents and for predicting flame extinction and extinguishment behaviors. Important findings of the study are summarized here.

Exit flow velocities, the various strain rates, and the stoichiometric scalar dissipation rates calculated for extinction of a given flame depend on the detailed chemical-kinetic mechanism used. This was clearly demonstrated for the methane flame. The mechanism employed here, GRI-Mech 1.2, predicts flames which are more robust than observed experimentally. On the other hand, the maximum flame temperatures for flames at extinction seem to be relatively independent of mechanism. This is relevant for the current work because a characteristic maximum flame temperature at extinction is used as the criteria for flame extinguishment.

Flame radiation is expected to have a noticeable effect on flame structure for the conditions relevant to extinguishment of diffusion flames. Recent detailed chemical-kinetic modeling investigations that incorporate radiation indicate that the relative importance of radiation increases with decreasing strain rate. For the strain rates identified as being typical of extinguishment of buoyancy-dominated diffusion flames, the effects of radiation represent a relatively small fraction of the total heat release. The neglect of radiation does not appear to seriously limit predictions of extinguishing concentrations, even though its neglect would be expected to change such properties as the extinguishing limit temperature and strain rate.

Assuming maximum flame temperatures at extinction of 1550 K for methane and 1433 K for propane as the extinguishing criteria provides good agreement between experimental results and the model predictions for the extinguishing volume fraction for nitrogen added to the air side of opposed-flow laminar diffusion flames stabilized on porous burners. Since the maximum flame temperatures at extinguishment seem to be relatively independent of mechanism, this finding suggests that there is a real difference between these two fuels. Predicted extinguishing concentrations are roughly 20 % greater for propane flames than for methane. This conclusion is consistent with the discussion of Maĉek who used experimental oxygen indices for various hydrocarbon fuels to predict that methane would have a higher extinguishing temperature and lower oxygen index than propane. [42] Interestingly, his analysis indicated that heptane flames should have flame temperatures very similar to those for methane and somewhat higher than for propane. This may explain why experimental cup burner measurements of extinguishment using thermal agents for heptane agree so closely with the detailed chemical-kinetic modeling results for the methane flame. A note of caution is required here since the amount of agent required to extinguish coflowing diffusion flames has been found to depend on burner diameter. The close agreement of the methane predictions and heptane measurements suggests that this effect is small for the cup burner. The current experimental results for the Santoro and Tsuji burners also indicate that propane flames require higher concentrations of a given thermal agent for extinguishment than the corresponding methane flames.

By assuming that the same limit temperatures derived from the nitrogen results could be applied to other agents, it was possible to obtain predicted extinguishing volume fractions of other thermal agents for methane and propane flames. An assessment of the validity of this hypothesis is complicated by the dependence of extinguishing concentrations for diffusion flames on burner type and configuration, however, it is clear from the results that the approach provides the correct ordering and relative magnitudes for a number of thermal agents (see Figure 38, Figure 48, and Figure 49). It is particularly noteworthy that the calculations correctly predict the experimental observation that nitrogen and helium have nearly equal extinguishing concentrations for both fuels. Other estimation approaches based simply on the heat capacities of the agents or adiabatic flame temperatures fail to successfully predict this experimental observation. For the methane flame the approach also provides an estimate for the extinguishing concentration of HFE7100 that is in excellent agreement with experimental estimates.

The calculations suggest that the relative effectiveness of thermal agents is the same for propane and methane flames despite the fact that the higher concentrations of a given agent are required to extinguish the propane flames. This conclusion is consistent with recent work of Babushok and Tsang, based on detailed chemical-kinetic modeling of premixed flames, that the relative effectiveness of agent does not vary with fuel. [70] Zegers et al. reached the same conclusion based on their experimental measurements using methanol, heptane, methane, and propane. [52]

Various approaches that have been used to estimate extinguishing concentrations of thermal agents in the past were discussed in Section III.A. One of these approaches was to identify the adiabatic flame temperature for a mixture of air and the thermal agent and to assume that this value would be a constant for other thermal agents. Figure 39 compares values of the calculated maximum flame temperature and the corresponding adiabatic flame temperatures for methane flames burning in air diluted with N_2, CO_2, Ar, He, and H_2O. The calculated maximum flame temperatures lie well below the corresponding adiabatic flame temperatures, but it is clear that, with the exception of helium, the adiabatic flame temperatures for extinguishing concentrations of the agents fall close together.

Table 18 lists adiabatic flame temperatures for both methane and propane flames burning in air for calculated extinguishing concentrations of N_2, CO_2, Ar, and He. For the three heavier agents the adiabatic flame temperatures for the extinguishing concentrations fall close together for both fuels, but particularly so for propane. The average adiabatic flame temperatures for the three agents are 1805 K and 1722 K for methane and propane, respectively. Based on results for these three agents one would conclude that the use of representative adiabatic flame temperature (fuel dependent) provides an effective approach for predicting extinguishing concentrations for a given fuel. Since earlier experimental studies often considered these three thermal agents, this was a reasonable conclusion. However, it is also clear from the results in Table 18 that the corresponding values for He, 1928 K for methane and 1903 K for propane, are much higher, reflecting an apparent higher effectiveness for this agent. Also recall that the adiabatic flame temperature corresponding to an extinguishing concentration of HFE7100 for the methane flame was 1696 K. This latter value suggests that HFE7100 is less effective than N_2, CO_2, and Ar as a thermal agent.

For a given adiabatic flame temperature, the amount of a thermal agent required should be directly proportional to its heat capacity. This is consistent with the early observation that extinguishing concentrations seemed to correlate inversely with heat capacity. However, as Larson [58] and Ewing et al. [60] have noted, this concept fails when extinguishing concentrations for the inert gases He (4.0 g/mol), Ne (20.2 g/mol), and Ar (39.9 g/mol) are compared (values in parentheses are atomic masses). These three gases have the same heat capacity, but the volume fraction required for extinguishment is ordered as He < Ne < Ar. These effects have been attributed to variations in thermal diffusivity, which increases as the atomic weight decreases.

The simple correlation of extinguishing concentration with heat capacity also disagrees with the finding reported here that simple dilution of the oxidizer by a species that does not absorb heat weakens the flame and can ultimately result in extinguishment. Since dilution effects increase as the agent concentration increases, an inverse relationship between the ability to extract heat and extinguishing concentration is not to be expected. The calculations run with surrogate agents designed to extract different known amounts of heat confirmed this expectation. Increasing the amount of heat extracted by a factor of two did not result in a factor of two decrease in the amount of agent required for extinguishment, but in a somewhat smaller change.

Thus there are at least three processes at work when a thermal agent extinguishes a flame, namely, absorption of heat, dilution, and removal of heat from the flame zone by thermal diffusion. Since removal of heat from the flame zone is favored by lower molecular weights, one would expect the effectiveness of N_2 to be somewhat increased relative to Ar and CO_2. The fact that that extinguishing concentrations are found to be inversely proportional to the heat capacities suggests that the effects of thermal conductivity and dilution roughly cancel for these species. The extinguishing concentration of HFE7100 is much lower than for N_2, CO_2, and Ar, with the result that the role of dilution is greatly reduced and a higher agent volume fraction is required for extinguishment than would be predicted based simply on thermal grounds. This leads to the lower adiabatic flame temperature calculated for the extinguishing condition for HFE 7100.

Zegers et al. have briefly discussed the relative roles of heat removal, dilution, and heat diffusion with regard to the effectiveness of thermal agents. [52] In particular, they argue that dilution and thermal conductivity effects are responsible for the fact that the effectiveness of high molecular weight thermal agents such as CF_4 and SF_6 are somewhat less than expected by comparison to such agents as N_2, CO_2, Ar, and He.

Several papers summarized in Section III.A have suggested that extinguishment by nearly all species can be understood simply in terms of their ability to extract heat. [58,59,60,61] These models require that some measure of the flame temperature at extinguishment, such as the maximum temperature or adiabatic flame temperature, depend on the flame system and agent used. On the other hand, the success of the current calculations in predicting extinguishment by thermal agents indicates that while the limit temperature is somewhat fuel dependent, the use of a single limit temperature is appropriate for different thermal agents. Furthermore, it was found that the location of heat extraction relative to the flame zone had no effect on the effectiveness of the agent. This contradicts one of the assumptions used in the correlation approach developed by Ewing et al. [61] Contrary to the conclusions of these earlier workers, we contend that variations of the limit temperature when different agents are used are actually an indication of the presence of chemical effects that weaken (or enhance) the flame and result in higher (or lower) limit temperatures. Such an effect can be observed in a recent detailed chemical-kinetic modeling study of flame inhibition by halon 1301. [72]

One of the findings of this investigation is that there is a substantial variation in the amount of a thermal agent required to extinguish a diffusion flame with burner type and, in the case of coflow burners, diameter. These differences were identified for methane, propane, and heptane flames. In general, opposed-flow flames were somewhat more difficult to extinguish than coflow flames attached to a burner. For the opposed-flow types the most stable flames appear to be those formed by having the fuel pass through a porous surface, while opposed-jet flames were somewhat easier to extinguish. Despite the similarity of the flames and flow conditions, a flame burning on a Santoro burner was considerably easier to extinguish than the corresponding cup burner flame. The reasons for these variations are not known at the present time. It is worthwhile to point out a major difference between the opposed-flow and coflow flames. Opposed-flow flames are formed near the stagnation plane of the opposed flows and are therefore truly stationary diffusion flames, with fuel and air mixing by molecular diffusion. In contrast, coflow flames are stabilized in the wake resulting from the no-slip boundary conditions near the burner exit at a location subject to a flow field. At the attachment point the flame is not a pure diffusion flame, but must propagate into a fuel/air mixture. Downstream of the attachment point the flame resembles a classical diffusion flame more closely. The existence of the propagating "edge flame" is probably responsible for the unusual flame behaviors observed during the Santoro Burner investigation. The success of the diffusion flame modeling in predicting the relative effects of agents on this burner suggests that the

inhibition of edge flames depends on thermal agent properties in much the same way as pure diffusion flames.

For this investigation we chose the most stable diffusion flames, i.e., opposed-flow porous burner type, as the basis for modeling and experimental comparison. This seems reasonable from a fire safety standpoint. It is worthwhile to point out that it should be possible to use the same calculational approach to predict extinguishing concentrations for the other burners, but the required limit temperatures will vary from flame to flame. The collapse of the data for different agents when plotted in terms of the burner exit velocities (e.g., see Figure 43) supports this contention.

VI. RECOMMENDATIONS FOR ADDITIONAL STUDY

HFE 7100 has been identified as a potential thermal-agent replacement for halon 1301. However, experiments in the TARPF were inconclusive. Additional screening tests in which the distribution of the released agent is more carefully controlled should be performed. Full-scale tests in simulated aircraft applications designed to assess the effectiveness of this agent should also be performed.

The tentative conclusion that agents released as liquids have higher than expected extinguishing efficiency than predicted based solely on their ability to extract heat and dilute the flame gases should be confirmed and investigated further to develop an understanding of the underlying mechanisms. It is possible that such an understanding would allow the design of more effective fire extinguishment systems.

In order to obtain a better understanding of droplet cooling of solid surfaces during fires, experiments performed under the conditions (droplet size, momentum, application rate, etc.) encountered in fire fighting are needed in the various heat transfer regimes. These types of experiments are still lacking in the literature. Surface cooling experiments involving liquid droplets and liquid surfaces are non-existent because the complexity of droplet-surface interaction makes the characterization of the heat transfer processes very difficult. A significant experimental effort in this area will be necessary to develop the knowledge and engineering correlations required for fire-fighting applications

VII. ACKNOWLEDGEMENTS

This research is part of the Department of Defense's Next Generation Program Fire Suppression Technology Program, funded in part by the DOD Strategic Environmental Research and Development Program under Contract W74RDV73243630.

VIII. APPENDIX A--DETAILED CHEMICAL-KINETIC MECHANISM FOR PROPANE COMBUSITON

Reactions are listed along with the three Arrhenius coefficients, A, β, E_a required to calculate the temperature-dependent rate constant, k, from

$$k = AT^\beta e^{-\frac{E_a}{RT}},$$

where units of A are in cgs units and vary with the order of reaction and E_a has units of cal/mol.

REACTION	A	β	E_a
H + H + M = H2 + M	1.00E18	-1.00	0.
H + H + H2 = 2H2	9.20E16	-0.60	0.
H + H + H2O = H2 + H2O	6.00E19	-1.25	0.
H + H + CO2 = H2 + CO2	5.49E20	-2.00	0.
2O + M = O2 + M	1.200E+17	-1.000	0.
H + OH + M = H2O + M	2.200E+22	-2.000	0.
O + H + M = OH + M	5.000E+17	-1.000	0.
O + OH = O2 + H	4.00E14	-0.50	0.
O + H2 = OH + H	5.06E04	2.67	6290.
2OH = O + H2O	1.50E09	1.14	99.
H2 + O2 = 2OH	1.70E13	0.00	47780.
OH + H2 = H2O + H	1.17E09	1.30	3626.
H + O2 + M = HO2 + M	3.61E17	-0.72	0.
H + HO2 = H2 + O2	1.25E13	0.00	0.
O + HO2 = O2 + OH	1.40E13	0.00	1073.
OH + HO2 = O2 + H2O	2.900E+13	0.000	-500.
2OH (+M) = H2O2 (+M)	7.400E+13	-0.370	0.
H + HO2 = 2OH	1.69E14	0.00	874.
HO2 + H = H2O + O	3.01E13	0.	1721.
2HO2 = O2 + H2O2	1.300E+11	0.000	-1630.00
H2O2 + H = H2 + HO2	4.79E13	0.	7950.
H2O2 + OH = H2O + HO2	1.00E13	0.00	1800.
O + H2O2 = OH + HO2	9.63E+6	2.000	4000.
CH4 + H = CH3 + H2	1.32E04	3.00	8040.
CH4 + O = CH3 + OH	1.02E09	1.50	8604.
CH4 + OH = CH3 + H2O	1.60E06	2.10	2460.
CH4 + O2 = CH3 + HO2	7.90E13	0.00	56000.
CH4 + HO2 = CH3 + H2O2	1.13E13	0.00	24641.
CH3 + H (+M) = CH4 (+M)	6.00E16	-1.00	0.
CH3 + H = CH2 + H2	9.00E13	0.00	15100.
CH3 + O = CH2O + H	8.00E13	0.00	0.
CH3 + OH = CH2 + H2O	7.50E06	2.00	5000.
CH3 + OH = CH3O + H	5.74E12	-0.23	13931.
CH3 + OH = CH2SING + H2O	8.90E19	-1.80	8067.

Reaction	A	n	E
CH3 + O2 = CH3O + O	2.05E18	-1.57	29229.
CH3 + O2 = CH2O + OH	3.59E09	-0.14	10150.
CH3 + HO2 = CH3O + OH	2.00E13	0.00	0.
CH3 + CH3 = C2H4 + H2	1.00E16	0.	32005.
CH3 + CH3 (+M) = C2H6 (+M)	9.03E16	-1.20	654.
H + CH2 (+M) = CH3 (+M)	2.500E+16	-0.800	0.00
CH2 + OH = CH2O + H	2.50E13	0.00	0.
CH2 + O = CO + 2H	5.00E13	0.00	0.
CH2 + CO2 = CH2O + CO	1.10E11	0.00	1000.
CH2 + O = CO + H2	3.00E13	0.00	0.
CH2 + O2 = CO2 + 2H	1.60E12	0.00	1000.
CH2 + O2 = CH2O + O	2.00E14	0.00	10000.
CH2 + O2 = CO2 + H2	6.90E11	0.00	500.
CH2 + O2 = CO + H2O	1.90E10	0.00	-1000.
CH2 + O2 = CO + OH + H	8.60E10	0.00	-500.
CH2 + O2 = HCO + OH	4.30E10	0.00	-500.
CH2 + CH3 = C2H4 + H	3.00E13	0.00	0.
2CH2 = C2H2 + H2	4.00E13	0.00	0.
CH2 + HO2 = CH2O + OH	3.01E13	0.	0.
CH2 + H2O2 = CH3O + OH	3.01E13	0.	0.
CH2 + CH2O = CH3 + HCO	1.20E12	0.	0.
CH2 + HCO = CH3 + CO	1.81E13	0.	0.
CH2SING + M = CH2 + M	1.00E13	0.00	0.
CH2SING + H2O (+M) = CH3OH (+M)	2.000E+13	0.000	0.00
CH2SING + H = CH2 + H	2.00E14	0.00	0.
O + CH2SING = H2 + CO	1.500E+13	0.000	0.00
O + CH2SING = H + HCO	1.500E+13	0.000	0.00
OH + CH2SING = H + CH2O	3.000E+13	0.000	0.00
CH2SING + CH3 = H + C2H4	1.80E13	0.	0.
CH2SING + O2 = H + OH + CO	2.800E+13	0.000	0.00
CH2SING + O2 = CO + H2O	1.200E+13	0.000	0.00
CH2SING + H2 = CH3 + H	7.00E13	0.00	0.
CH2SING + CH2O = CH3 + HCO	4.00E13	0.	0.
CH2SING + CO2 = CO + CH2O	1.00E+13	0.	0.
CH2SING + CH4 = 2CH3	4.00E13	0.00	0.
CH2SING + C2H6 = CH3 + C2H5	1.20E14	0.00	0.
CH2 + H = CH + H2	1.00E18	-1.56	0.
CH2 + OH = CH + H2O	1.13E07	2.00	3000.
CH + O2 = HCO + O	3.30E13	0.00	0.
CH + O = CO + H	5.70E13	0.00	0.
H + CH = C + H2	1.1E+14	0.00	0.
CH + OH = HCO + H	3.00E13	0.00	0.
CH + CO2 = HCO + CO	3.40E12	0.00	690.
CH + H2O = CH2O + H	1.17E15	-0.75	0.
CH + CH2O = CH2CO + H	9.46E13	0.00	-515.
CH + CH2 = C2H2 + H	4.00E13	0.00	0.

Reaction	A	n	E
CH + CH3 = C2H3 + H	3.00E13	0.00	0.
CH + CH4 = C2H4 + H	6.00E13	0.00	0.
C2H3 + CH = CH2 + C2H2	5.00E13	0.00	0.
HCCO + CH = C2H2 + CO	5.00E13	0.00	0.
CH + CO (+M) = HCCO (+M)	5.000E+13	0.000	0.00
OH + C = H + CO	5.000E+13	0.000	0.00
C + O2 = O + CO	5.800E+13	0.000	576.00
C + CH2 = H + C2H	5.000E+13	0.000	0.00
C + CH3 = H + C2H2	5.000E+13	0.000	0.00
CH3OH + H = CH3O + H2	4.24E06	2.11	4870.
CH3OH + H = CH2OH + H2	1.7E07	2.11	4870.
CH3OH + OH = CH3O + H2O	6.00E13	0.00	5500.
CH3OH + OH = CH2OH + H2O	7.23E12	0.00	1610.
CH3OH + HO2 = CH3O + H2O2	1.00E12	0.00	18400.
CH3OH + CH3 = CH3O + CH4	1.45E+1	3.10	6935.
CH3OH + O = CH2OH + OH	3.88E05	2.50	3080.
O + CH3OH = OH + CH3O	1.300E+05	2.50	5000.
CH3OH + HO2 = CH2OH + H2O2	1.0E12	0.00	10000.
CH3OH + O2 = CH2OH + HO2	2.05E13	0.00	44910.
CH3OH + CH3 = CH2OH + CH4	3.20E01	3.20	7170.
CH3OH + CH3O = CH3OH + CH2OH	6.E11	0.	4070.
CH3OH + C2H5 = C2H6 + CH2OH	31.9	3.2	9160.
CH3OH + C2H5 = C2H6 + CH3O	14.4	3.1	8940.
CH3OH + CH3CO = CH3HCO + CH2OH	4.85E3	3.	12340.
CH3OH + C2H3 = C2H4 + CH2OH	31.9	3.2	7170.
CH3OH + C2H3 = C2H4 + CH3O	14.4	3.1	6940.
CH3OH + CH2 = CH3 + CH2OH	31.9	3.2	7170.
CH3OH + CH2 = CH3 + CH3O	14.4	3.1	6940.
CH3OH (+M) = CH3 + OH (+M)	1.90E16	0.	91780.
CH3OH (+M) = CH2OH + H (+M)	1.54E16	0.	96844.
CH3CO + M = CH3 + CO + M	1.20E15	0.	12516.
CH3CO + H = CH3 + HCO	9.64E13	0.	0.
CH3CO + O = CH3 + CO2	9.64E12	0.	0.
CH3CO + OH = CH2CO + H2O	1.20E13	0.	0.
CH3CO + OH = CH3 + CO + OH	3.01E13	0.	0.
CH3CO + HO2 = CH3 + CO2 + OH	3.01E13	0.	0.
CH3CO + HCO = CH3HCO + CO	9.03E12	0.	0.
CH3CO + C2H6 = C2H5 + CH3HCO	1.81E04	2.75	17517.
CH3CO + CH2O = CH3HCO + HCO	1.81E11	0.	12909.
CH3O + M = CH2O + H + M	1.00E14	0.00	25000.
CH3O + O2 = CH2O + HO2	6.30E10	0.00	2600.
CH3O + H = CH2O + H2	2.00E13	0.00	0.
H + CH3O = H + CH2OH	3.400E+06	1.600	0.00
H + CH3O = CH2SING + H2O	1.600E+13	0.000	0.00
H + CH3O (+M) = CH3OH (+M)	5.000E+13	0.000	0.00
CH3O + O = CH2O + OH	1.00E13	0.00	0.

Reaction	A	n	E
CH3O + OH = CH2O + H2O	1.00E13	0.00	0.
CH3O + HO2 = CH2O + H2O2	3.01E11	0.	0.
CH3O + CO = CH3 + CO2	1.57E13	0.	11797.
CH3O + C2H5 = CH2O + C2H6	2.41E13	0.	0.
CH3O + C2H3 = CH2O + C2H4	2.41E13	0.	0.
CH3O + C2H = CH2O + C2H2	2.41E13	0.	0.
CH3O + CH3 = CH4 + CH2O	2.40E13	0.	0.
CH2OH + M = CH2O + H + M	1.00E14	0.00	25000.
CH2OH + O2 = CH2O + HO2	2.41E14	0.00	5000.
CH2OH + H = CH3 + OH	9.64E13	0.	0.
CH2OH + H = CH2O + H2	6.03E12	0.	0.
CH2OH + HO2 = CH2O + H2O2	1.20E13	0.	0.
CH2OH + HCO = CH3OH + CO	1.20E14	0.	0.
CH2OH + HCO = CH2O + CH2O	1.81E14	0.	0.
CH2OH + CH3 = C2H5 + OH	1.37E14	-.41	6589.
CH2OH + CH2O = HCO + CH3OH	5.54E03	2.81	5862.
CH2OH + CH2OH = CH3OH + CH2O	1.20E13	0.	0.
CH2OH + O = CH2O + OH	2.40E13	0.	0.
CH2O + M = HCO + H + M	3.31E16	0.00	81000.
CH2O + H = HCO + H2	2.19E08	1.77	3000.
CH2O + O = HCO + OH	1.80E13	0.00	3080.
CH2O + OH = HCO + H2O	3.43E09	1.18	-447.
CH2O + HO2 = HCO + H2O2	1.99E12	0.00	11665.
CH2O + O2 = HCO + HO2	2.04E13	0.00	38900.
CH2O + CH3 = HCO + CH4	4.09E12	0.00	8843.
H2 + CO (+M) = CH2O (+M)	4.300E+07	1.500	79600.00
HCO + M = CO + H + M	2.50E14	0.00	16802.
HCO + H = CO + H2	1.19E13	0.25	0.
HCO + O = CO + OH	3.00E13	0.00	0.
HCO + O = CO2 + H	3.00E13	0.00	0.
HCO + OH = CO + H2O	1.00E14	0.00	0.
HCO + O2 = HO2 + CO	7.600E+12	0.000	400.00
HCO + HO2 = CO2 + OH + H	3.00E13	0.	0.
CH3 + HCO = CH4 + CO	1.20E14	0.	0.
O + CO + M = CO2 + M	6.020E+14	0.00	3000.
CO + OH = CO2 + H	1.51E07	1.30	-758.
O2 + CO = O + CO2	2.500E+12	0.00	47800.
CO + HO2 = CO2 + OH	5.80E13	0.00	22934.
C2H6 + H = C2H5 + H2	5.40E02	3.50	5210.
C2H6 + O = C2H5 + OH	3.00E07	2.00	5115.
C2H6 + OH = C2H5 + H2O	8.70E09	1.05	1810.
C2H6 + CH3 = C2H5 + CH4	5.50E-1	4.00	8300.
C2H6 + O2 = C2H5 + HO2	4.03E13	0.	50842.
C2H6 + HO2 = C2H5 + H2O2	2.95E11	0.	14935.
H + C2H5 (+M) = C2H6 (+M)	5.21E+17	-0.990	1580.
C2H5 + H = CH3+CH3	1.00E14	0.00	0.

Reaction	A	n	E
C2H5 + H = C2H4 + H2	1.81E12	0.	0.
C2H5 + O = CH3HCO + H	5.50E13	0.	0.
C2H5 + O = CH2O + CH3	1.10E13	0.	0.
C2H5 + O2 = C2H4 + HO2	2.56E19	-2.77	1977.
C2H5 + OH = C2H4 + H2O	2.41E13	0.	0.
C2H5 + HO2 = CH3 + CH2O + OH	2.40E13	0.	0.
C2H4 + M = C2H2 + H2 + M	1.50E15	0.00	55800.
C2H4 + M = C2H3 + H + M	1.40E16	0.00	82360.
H + C2H4 (+M) = C2H5 (+M)	1.080E+12	0.454	1820.00
H + C2H4 = C2H3 + H2	1.325E+06	2.530	12240.00
C2H4 + O = CH3 + HCO	1.60E09	1.20	746.
C2H4 + OH = C2H3 + H2O	4.50E06	2.00	2850.
CH3 + C2H4 = C2H3 + CH4	2.270E+05	2.000	9200.
C2H4 + O2 = C2H3 + HO2	4.22E13	0.	57594.
C2H4 + CO = C2H3 + HCO	1.51E14	0.	90562.
C2H3 + H = C2H2 + H2	1.20E13	0.00	0.
C2H3 + OH = C2H2 + H2O	5.00E12	0.00	0.
C2H3 + CH2 = C2H2 + CH3	3.00E13	0.00	0.
C2H3 + O2 = CH2O + HCO	1.05E38	-8.22	7030.
C2H3 + O = CH2CO + H	3.00E13	0.00	0.
C2H3 + O2 = C2H2 + HO2	1.20E11	0.	0.
C2H3 + HO2 = CH2CO + OH + H	3.00E13	0.	0.
C2H2 + H (+M) = C2H3 (+M)	5.54E12	0.00	2410.
C2H2 + OH = HCCOH + H	5.04E05	2.30	13500.
C2H2 + OH = CH2CO + H	2.18E-4	4.50	-1000.
C2H2 + OH = CH3 + CO	4.83E-4	4.00	-2000.
C2H2 + O = CH2 + CO	1.02E07	2.00	1900.
C2H2 + O = HCCO + H	1.02E07	2.00	1900.
O + C2H2 = OH + C2H	4.600E+19	-1.410	28950.
C2H2 + O2 = HCCO + OH	2.00E08	1.50	30100.
C2H2 = C2H + H	1.80E41	-7.76	137510.
C2H2 + H = C2H + H2	6.02E13	0.	22243.
C2H2 + OH = C2H + H2O	1.45E4	2.68	12035.
C2H2 + O2 = C2H + HO2	1.20E13	0.	74475.
C2H + O = CH + CO	1.81E13	0.	0.
OH + C2H = H + HCCO	2.000E+13	0.	0.
OH + C2H = CH2 + CO	2.00E13	0.	0.
C2H + O2 = CO + HCO	2.41E12	0.	0.
HCCOH + H = CH2CO + H	1.00E13	0.00	0.
CH2CO + H = CH3 + CO	1.13E13	0.00	3428.
CH2CO + H = HCCO + H2	5.00E13	0.00	8000.
CH2CO + O = CO2 + CH2	1.75E12	0.00	1350.
CH2CO + O = HCCO + OH	1.00E13	0.00	8000.
CH2CO + OH = HCCO + H2O	7.50E12	0.00	2000.
CH2CO (+M) = CH2 + CO (+M)	3.00E14	0.00	70980.
CH2CO + O = HCO + HCO	2.00E13	0.	2293.

Reaction	A	n	E
CH2CO + O = CH2O + CO	2.00E13	0.	0.
CH2CO + OH = CH2O + HCO	2.80E13	0.	0.
HCCO + OH = HCO + CO + H	1.00E13	0.	0.
HCCO + CH2 = C2H + CH2O	1.00E13	0.	2000.
HCCO + H = CH2SING + CO	1.00E14	0.00	0.
HCCO + O = H + 2CO	1.00E14	0.00	0.
HCCO + O2 = 2CO + OH	1.60E12	0.00	854.
2HCCO = C2H2 + 2CO	1.00E13	0.00	0.
HCCO + CH2 = C2H3 + CO	3.00E13	0.00	0.
CH2 + C2H2 = H2CCCH + H	1.20E+13	0.0	6600.0
CH2SING + C2H2 = H2CCCH + H	1.50E+14	0.0	0.0
CH2SING + C2H4 = aC3H5 + H	1.30E+14	0.0	0.0
CH + C2H2 = C3H2 + H	1.00E+14	0.0	0.0
CH3CH2O + M = CH3HCO + H + M	1.16E+35	-5.89	25274.0
CH3CH2O + M = CH3 + CH2O + M	1.35E+38	-6.96	23800.0
CH3CH2O + O2 = CH3HCO + HO2	4.68E+2	3.16	5380.0
CH3CH2O + H = CH3 + CH2OH	1.00E+13	0.0	0.0
CH3CH2O + H = CH3HCO + H2	1.00E+13	0.0	0.0
CH3CH2O + OH = CH3HCO + H2O	1.00E+13	0.0	0.0
CH3CHOH + O2 = CH3HCO + HO2	4.82E+14	0.0	5017.0
CH3CHOH + CH3 = iC3H7 + OH	2.00E+13	0.0	0.0
CH3CHOH + O = CH3HCO + OH	1.00E+14	0.0	0.0
CH3CHOH + H = CH3HCO + H2	1.00E+13	0.0	0.0
CH3CHOH + H = CH3 + CH2OH	1.00E+13	0.0	0.0
CH3CHOH + HO2 = CH3HCO + OH + OH	4.00E+13	0.0	0.0
CH3CHOH + OH = CH3HCO + H2O	5.00E+12	0.0	0.0
CH3CHOH + M = CH3HCO + H + M	1.00E+14	0.0	25000.0
CH3HCO + OH = CH3CO + H2O	9.24E+6	1.5	-962.0
CH3HCO + OH = CH2HCO + H2O	1.72E+5	2.4	815.0
CH3HCO + O = CH3CO + OH	1.77E+18	-1.9	2975.0
CH3HCO + O = CH2HCO + OH	3.72E+13	-0.2	3556.0
CH3HCO + H = CH3CO + H2	4.66E+13	-0.35	2988.0
CH3HCO + H = CH2HCO + H2	1.85E+12	0.40	5359.0
CH3HCO + CH3 = CH3CO + CH4	3.90E-7	5.8	2200.0
CH3HCO + CH3 = CH2HCO + CH4	2.45E+1	3.15	5727.0
CH3HCO + HO2 = CH3CO + H2O2	2.40E+19	-2.2	14030.0
CH3HCO + HO2 = CH2HCO + H2O2	2.32E+11	0.40	14864.0
CH3HCO + O2 = CH3CO + HO2	1.00E+14	0.00	42200.0
C2H5 + HO2 = CH3CH2O + OH	3.00E+13	0.0	0.0
C2H5 + O2 = CH3HCO + OH	4.90E+11	-0.48	8357.0
C2H4 + OH = C2H4OH	1.29E+12	0.0	-817.0
C2H4OH + O2 = HOC2H4O2	1.00E+12	0.0	-1100.0
HOC2H4O2 = CH2O + CH2O + OH	6.00E+10	0.0	24500.0
C2H4 + O = CH2HCO + H	3.39E+06	1.88	179.0
C2H3 + O2 = CH2HCO + O	5.50E+14	-0.611	5260.0
C2H3 + CH3 = aC3H5 + H	4.73E+02	3.7	5677.

Reaction	A	n	E
C2H3 + CH3 = C3H6	4.46E+56	-13.0	13865.0
CH2HCO + H = CH3 + HCO	5.00E+13	0.0	0.0
CH2HCO + H = CH2CO + H2	2.00E+13	0.0	0.0
CH2HCO + O = CH2O + HCO	1.00E+14	0.0	0.0
CH2HCO + OH = CH2CO + H2O	3.00E+13	0.0	0.0
CH2HCO + O2 = CH2O + CO + OH	3.00E+10	0.0	0.0
CH2HCO + CH3 = C2H5 + CO + H	4.90E+14	-0.50	0.0
CH2HCO + HO2 = CH2O + HCO + OH	7.00E+12	0.0	0.0
CH2HCO + HO2 = CH3HCO + O2	3.00E+12	0.0	0.0
CH2HCO = CH3 + CO	1.17E+43	-9.83	43756.0
CH2HCO = CH2CO + H	1.81E+43	-9.61	45868.0
CHOCHO (+M) = CH2O + CO (+M)	4.27E+12	0.0	50600.0
CHOCHO = CO + CO + H2	4.07E+42	-8.5	69278.0
CHOCHO + OH = HCO + CO + H2O	1.00E+13	0.0	0.0
CHOCHO + O = HCO + CO + OH	7.24E+12	0.0	1970.0
CHOCHO + H = CH2O + HCO	1.00E+12	0.0	0.0
CHOCHO + HO2 = HCO + CO + H2O2	1.70E+12	0.0	10700.0
CHOCHO + CH3 = HCO + CO + CH4	1.74E+12	0.0	8440.0
CHOCHO + O2 = HCO + CO + HO2	1.00E+14	0.0	37000.0
HCCO + C2H2 = H2CCCH + CO	1.00E+11	0.0	3000.0
HCCO + OH = C2O + H2O	3.00E+13	0.0	0.0
C2O + H = CH + CO	1.00E+13	0.0	0.0
C2O + O = CO + CO	5.00E+13	0.0	0.0
C2O + OH = CO + CO + H	2.00E+13	0.0	0.0
C2O + O2 = CO + CO + O	2.00E+13	0.0	0.0
C3H8 (+M) = C2H5 + CH3 (+M)	7.90E+22	-1.8	88629.0
C3H8 + O2 = iC3H7 + HO2	4.00E+13	0.0	48610.0
C3H8 + O2 = nC3H7 + HO2	4.00E+13	0.0	51360.0
C3H8 + HO2 = nC3H7 + H2O2	4.76E+04	2.55	16492.0
C3H8 + HO2 = iC3H7 + H2O2	9.64E+03	2.6	13909.0
C3H8 + OH = nC3H7 + H2O	3.16E+07	1.8	934.0
C3H8 + OH = iC3H7 + H2O	7.08E+06	1.9	-159.0
C3H8 + O = nC3H7 + OH	3.73E+06	2.4	5504.0
C3H8 + O = iC3H7 + OH	5.48E+05	2.5	3139.0
C3H8 + H = iC3H7 + H2	1.30E+06	2.4	4471.0
C3H8 + H = nC3H7 + H2	1.33E+06	2.54	6756.0
C3H8 + CH3 = nC3H7 + CH4	9.04E-01	3.65	7153.0
C3H8 + CH3 = iC3H7 + CH4	1.51E+00	3.46	5480.0
C3H8 + C2H3 = iC3H7 + C2H4	1.00E+03	3.1	8830.0
C3H8 + C2H3 = nC3H7 + C2H4	6.00E+02	3.3	10500.0
C3H8 + C2H5 = iC3H7 + C2H6	1.51E+00	3.46	7470.0
C3H8 + C2H5 = nC3H7 + C2H6	9.03E-01	3.65	9140.0
C3H8 + aC3H5 = C3H6 + nC3H7	2.35E+02	3.3	19842.0
C3H8 + aC3H5 = C3H6 + iC3H7	7.83E+01	3.3	18169.0
nC3H7 (+M) = C2H4 + CH3 (+M)	1.23E+13	-0.1	30202.0
nC3H7 + O2 = C3H6 + HO2	3.58E+09	0.0	-3532.0

Reaction	A	n	Ea
iC3H7 + O2 = C3H6 + HO2	6.10E+20	-2.86	7910.0
C3H6 + H (+M) = iC3H7 (+M)	5.70E+09	1.16	874.0
iC3H7 + H = C2H5 + CH3	5.00E+13	0.0	0.0
nC3H7 + H = C2H5 + CH3	1.00E+14	0.0	0.0
pC3H5 + H = C3H6	7.0e13	0.0	0.0
sC3H5 + H = C3H6	5.0e13	0.0	0.0
C3H6 = C2H2 + CH4	2.50E+12	0.0	70000.0
C3H6 = aC3H4 + H2	3.00E+13	0.0	80000.0
C3H6 + HO2 = aC3H5 + H2O2	9.64E+03	2.6	13910.0
C3H6 + OH + O2 = CH3HCO + CH2O + OH	3.00E+10	0.0	-8280.0
C3H6 + OH = aC3H5 + H2O	3.12E+06	2.0	-298.0
C3H6 + OH = sC3H5 + H2O	1.11E+06	2.0	1451.0
C3H6 + OH = pC3H5 + H2O	2.11E+06	2.0	2778.0
C3H6 + O = CH3CHCO + H + H	5.01E+07	1.76	76.0
C3H6 + O = C2H5 + HCO	1.58E+07	1.76	-1216.0
C3H6 + O = aC3H5 + OH	5.24E+11	0.7	5884.0
C3H6 + O = pC3H5 + OH	1.20E+11	0.7	8959.0
C3H6 + O = sC3H5 + OH	6.03E+10	0.7	7632.0
C3H6 + H = C2H4 + CH3	7.23E+12	0.0	1302.0
C3H6 + H = aC3H5 + H2	1.73E+05	2.5	2492.0
C3H6 + H = sC3H5 + H2	4.09E+05	2.5	9794.0
C3H6 + H = pC3H5 + H2	8.04E+05	2.5	12284.0
aC3H5 + HO2 = C3H6 + O2	3.0e12	0.0	0.0
C3H6 + CH3 = aC3H5 + CH4	2.22E+00	3.5	5675.0
C3H6 + CH3 = sC3H5 + CH4	8.43E-01	3.5	11656.0
C3H6 + CH3 = pC3H5 + CH4	1.35E+00	3.5	12848.0
C3H6 + HCO = aC3H5 + CH2O	1.08E+07	1.9	17010.0
CH3CHCO + OH = CH2CHCO + H2O	4.00E+06	2.0	0.0
CH3CHCO + O = CH2CHCO + OH	7.60E+08	1.5	8500.0
CH3CHCO + H = CH2CHCO + H2	2.00E+05	2.5	2500.0
CH3CHCO + H = C2H5 + CO	2.00E+13	0.0	2000.0
CH3CHCO + O = CH3 + HCO + CO	3.00E+07	2.0	0.0
CH2CHCHO + OH = CH2CHCO + H2O	1.00E+13	0.0	0.0
CH2CHCHO + O = CH2CHCO + OH	7.24E+12	0.0	1970.0
CH2CHCHO + O = CH2CO + HCO + H	5.01E+07	1.76	76.0
CH2CHCHO + H = CH2CHCO + H2	3.98E+13	0.0	4200.0
CH2CHCHO + H = C2H4 + HCO	2.00E+13	0.0	3500.0
CH2CHCHO + O2 = CH2CHCO + HO2	3.00E+13	0.0	36000.0
CH2CHCO = C2H3 + CO	1.00E+14	0.0	34000.0
CH2CHCO + O = C2H3 + CO2	1.00E+14	0.0	0.0
aC3H5 + O2 = CH2CHCHO + OH	1.82E+13	-0.41	22859.0
aC3H5 + O2 = aC3H4 + HO2	4.99E+15	-1.4	22428.0
aC3H5 + O2 = CH2HCO + CH2O	1.06E+10	0.34	12838.0
aC3H5 + O2 = C2H2 + CH2O + OH	2.78E+25	-4.8	15468.0
aC3H5 + HO2 = CH2CHCHO + H + OH	1.00E+13	0.0	0.0
aC3H5 + OH = aC3H4 + H2O	1.00E+13	0.0	0.0

aC3H5 + H = aC3H4 + H2	5.00E+13	0.0	0.0
aC3H5 + H = C3H6	1.88E+26	-3.6	5468.0
aC3H5 + O = CH2CHCHO + H	1.81E+14	0.0	0.0
aC3H5 + CH3 = aC3H4 + CH4	3.02E+12	-0.32	-131.0
aC3H5 + CH3 = C4H8-1	1.76E+50	-11.0	18600.0
pC3H5 + O2 = CH3HCO + HCO	1.09E+23	-3.29	3892.0
pC3H5 + O2 = CH3CHCO + H + O	1.60E+15	-0.78	3135.0
pC3H5 + O = CH3CHCO + H	1.00E+14	0.0	0.0
pC3H5 + H = pC3H4 + H2	2.00E+13	0.0	0.0
pC3H5 + OH = pC3H4 + H2O	1.00E+13	0.0	0.0
pC3H5 + H = aC3H5 + H	1.00E+14	0.0	0.0
sC3H5 + H = aC3H5 + H	1.00E+14	0.0	0.0
sC3H5 + O2 = CH3CO + CH2O	1.09E+22	-3.29	3892.0
sC3H5 + O = CH2CO + CH3	1.00E+14	0.0	0.0
sC3H5 + H = pC3H4 + H2	4.00E+13	0.0	0.0
sC3H5 + OH = pC3H4 + H2O	2.00E+13	0.0	0.0
aC3H4 + H = H2CCCH + H2	2.0e7	2.0	5000.
aC3H4 + O = C2H4 + CO	1.34E+07	1.88	179.0
aC3H4 + OH = H2CCCH + H2O	1.00e7	2.0	1000.0
aC3H4 + CH3 = H2CCCH + CH4	1.5e0	3.5	5600.
aC3H4 = pC3H4	1.48e13	0.0	60401.
pC3H4 + H = H2CCCH + H2	2.00e7	2.0	5000.
pC3H4 + O = C2H4 + CO	1.50E+13	0.0	2102.0
pC3H4 + OH = H2CCCH + H2O	1.00e7	2.0	1000.0
pC3H4 + CH3 = H2CCCH + CH4	1.5e0	3.5	5600.
pC3H4 + H = CH3 + C2H2	5.12e10	1.0	2060.
pC3H4 + H (+M) = sC3H5 (+M)	6.50E+12	0.0	2000.0
aC3H4 + H (+M) = aC3H5 (+M)	1.20E+11	0.69	3007.0
aC3H4 + H (+M)= sC3H5 (+M)	8.49E+12	0.0	2000.0
H2CCCH + O2 = CH2CO + HCO	3.00E+10	0.0	2868.0
H2CCCH + O = CH2O + C2H	2.00E+13	0.0	0.0
H2CCCH + H = C3H2 + H2	5.00E+13	0.0	3000.0
H2CCCH + OH = C3H2 + H2O	2.00E+13	0.0	0.0
H2CCCH + CH3 = CH3CHCCH2	5.00E+12	0.0	0.0
H2CCCH + CH3 = CH3CH2CCH	5.00E+12	0.0	0.0
H2CCCH + CH = HCCHCCH + H	7.00E+13	0.0	0.0
H2CCCH + CH = H2CCCCH + H	7.00E+13	0.0	0.0
H2CCCH + H (+M) = aC3H4 (+M)	1.66E+15	-0.37	0.0
H2CCCH + H (+M) = pC3H4 (+M)	1.66E+15	-0.37	0.0
C3H2 + O2 = HCCO + CO + H	5.00E+13	0.0	0.0
C3H2 + OH = C2H2 + HCO	5.00E+13	0.0	0.0
CHCHCHO + O2 = C2H2 + CO + HO2	3.00E+12	0.0	0.0
CHCHCHO = C2H2 + HCO	1.00E+14	0.0	33000.0
CHCHCHO + H = CH2CHCO + H	1.00E+14	0.0	0.0
CHCHCHO + OH = HCCCHO + H2O	1.00E+13	0.0	0.0
CHCHCHO + H = HCCCHO + H2	2.00E+13	0.0	0.0

Reaction	A	n	E
HCCCHO + H = C2H2 + HCO	1.00E+14	0.0	3000.0
HCCCHO + OH = HCCCO + H2O	1.00E+13	0.0	0.0
HCCCHO + H = HCCCO + H2	4.00E+13	0.0	4200.0
HCCCO + O2 = HCO + CO + CO	1.40E+09	1.0	0.0
HCCCO + H = C2H2 + CO	1.00E+14	0.0	0.0
C4H10 = C2H5 + C2H5	2.00E+16	0.0	81300.0
C4H10 = nC3H7 + CH3	1.74E+17	0.0	85700.0
C4H10 = pC4H9 + H	1.00E+14	0.0	100000.0
C4H10 = sC4H9 + H	1.00E+14	0.0	100000.0
C4H10 + O2 = pC4H9 + HO2	2.50E+13	0.0	49000.0
C4H10 + O2 = sC4H9 + HO2	4.00E+13	0.0	47600.0
C4H10 + aC3H5 = pC4H9 + C3H6	7.94E+11	0.0	20500.0
C4H10 + aC3H5 = sC4H9 + C3H6	3.16E+11	0.0	16400.0
C4H10 + CH3 = pC4H9 + CH4	5.0e11	0.0	13600.0
C4H10 + CH3 = sC4H9 + CH4	4.3e11	0.0	10500.0
C4H10 + H = pC4H9 + H2	2.84e5	2.54	6050.
C4H10 + H = sC4H9 + H2	4.13E+07	1.73	753.0
C4H10 + OH = sC4H9 + H2O	7.23E+07	1.64	-247.0
C4H10 + O = pC4H9 + OH	1.13E+14	0.0	7850.0
C4H10 + O = sC4H9 + OH	5.62E+13	0.0	5200.0
C4H10 + HO2 = pC4H9 + H2O2	1.70E+13	0.0	20460.0
C4H10 + HO2 = sC4H9 + H2O2	1.12E+13	0.0	17700.0
sC4H9 (+M) = C3H6 + CH3 (+M)	2.14e12	0.65	30856.
sC4H9 = C4H8-1 + H	2.00E+13	0.0	40400.0
sC4H9 = C4H8-2 + H	5.01E+12	0.0	37900.0
pC4H9 (+M) = C2H5 + C2H4 (+M)	1.06e13	0.0	27828.0
pC4H9 = C4H8-1 + H	1.26E+13	0.0	38600.0
C4H8-1 = C2H3 + C2H5	1.00E+19	-1.0	96770.0
C4H8-1 = H + C4H7	4.11E+18	-1.0	97350.0
C4H8-1 + CH3 = C4H7 + CH4	1.00E+11	0.0	7300.0
C4H8-1 + H = C4H7 + H2	5.00E+13	0.0	3900.0
C4H8-1 + O = nC3H7 + HCO	1.80E+05	2.5	-1029.0
C4H8-1 + O = CH2CHCHO + CH3 + H	9.67E+04	2.5	-1029.0
C4H8-1 + OH = C4H7 + H2O	2.25E+13	0.0	2217.0
C4H8-1 + aC3H5 = C4H7 + C3H6	7.90E+10	0.0	12400.0
C4H8-1 + O2 = C4H7 + HO2	4.00E+12	0.0	33200.0
C4H8-2 = H + C4H7	4.11E+18	-1.0	97350.0
C4H8-2 + CH3 = C4H7 + CH4	1.00E+11	0.0	8200.0
C4H8-2 + H = C4H7 + H2	5.00E+13	0.0	3800.0
C4H8-2 + O = iC3H7 + HCO	2.79E+06	2.12	-1775.0
C4H8-2 + OH = C4H7 + H2O	3.90E+13	0.0	2217.0
C4H8-2 + O = CH3CO + C2H5	1.53E+07	1.87	-1476.0
C4H8-2 + O = CH3 + CH3CHCO + H	8.22E+06	1.87	-1476.0
C4H8-2 + O2 = C4H7 + HO2	8.00E+13	0.0	37400.0
C4H7 = CH2CHCHCH2 + H	1.00E+14	0.0	55000.0
C4H7 + OH = CH2CHCHCH2 + H2O	1.00E+13	0.0	0.0

Reaction	A	n	E
C4H7 + CH3 = CH2CHCHCH2 + CH4	8.00E+12	0.0	0.0
C4H7 + aC3H5 = C3H6 + CH2CHCHCH2	6.31E+12	0.0	0.0
C4H7 + O2 = CH2CHCHCH2 + HO2	1.00E+09	0.0	0.0
C4H7 + H = CH2CHCHCH2 + H2	3.16E+13	0.0	0.0
CH2CHCHCH2 + OH = CH2CHCHCH + H2O	2.00E+07	2.0	5000.0
CH2CHCHCH2 + OH = CH2CHCCH2 + H2O	2.00E+07	2.0	2000.0
CH2CHCHCH2 + O = HCO + aC3H5	6.02E+08	1.45	-858.0
CH2CHCHCH2 + O = CH2HCO + C2H3	1.00E+12	0.0	0.0
CH2CHCHCH2 + H = CH2CHCHCH + H2	3.00E+07	2.0	13000.0
CH2CHCHCH2 + H = CH2CHCCH2 + H2	3.00E+07	2.0	6000.0
CH3CH2CCH + OH = CH3CHCCH + H2O	1.00E+07	2.0	2000.0
CH3CH2CCH + H = C2H5 + C2H2	1.00E+14	0.0	3000.0
CH3CHCCH2 + OH = CH2CHCCH2 + H2O	2.00E+07	2.0	1000.0
CH3CHCCH2 + OH = CH3CCCH2 + H2O	1.00E+07	2.0	2000.0
CH3CHCCH2 + OH = CH3CHCCH + H2O	2.00E+07	2.0	2500.0
CH3CHCCH2 + H = CH2CHCCH2 + H2	5.00E+07	2.0	5000.0
CH3CHCCH2 + H = CH3CCCH2 + H2	1.50E+07	2.0	6000.0
CH3CHCCH2 + H = CH3CHCCH + H2	3.00E+07	2.0	6500.0
CH3CHCCH2 + H = CH3 + aC3H4	2.00E+13	0.0	2000.0
CH3CHCCH + H = CH3 + H2CCCH	1.00E+14	0.0	0.0
CH3CHCCH + O2 = CH3CHCO + HCO	4.16E+10	0.0	2510.0
CH3CHCCH + OH = CH2CHCCH + H2O	3.00E+13	0.0	0.0
CH2CHCCH2 + H = CH3 + H2CCCH	1.00E+14	0.0	0.0
CH2CHCCH2 + H = CH3CCCH2 + H	3.00E+13	0.0	0.0
CH3CCCH2 + H = CH3 + H2CCCH	1.00E+14	0.0	0.0
CH3CCCH2 + O2 = CH3CO + CH2CO	4.16E+10	0.0	2510.0
CH3CCCH2 + H = H2CCCCH2 + H2	1.00E+14	0.0	8000.0
CH3CCCH2 + OH = H2CCCCH2 + H2O	1.00E+13	0.0	0.0
CH2CHCHCH + H = CH2CHCCH2 + H	1.0E+14	0.0	0.0
CH2CHCHCH + OH = CH2CHCCH + H2O	2.00E+07	2.0	1000.0
CH2CHCHCH + H = CH2CHCCH + H2	3.00E+07	2.0	1000.0
CH3CHCCH (+M) = CH2CHCCH + H (+M)	1.00E+13	0.0	49000.0
CH3CCCH2 (+M) = H2CCCCH2 + H (+M)	1.00E+13	0.0	56000.0
CH2CHCCH2 (+M) = CH2CHCCH + H (+M)	1.00E+14	0.0	50000.0
CH2CHCHCH (+M) = CH2CHCCH + H (+M)	1.00E+14	0.0	37000.0
CH2CHCHCH + O2 = CHCHCHO + CH2O	1.00E+12	0.0	0.0
CH2CHCHCH + O2 = CH2CHCCH + HO2	1.00E+07	2.0	10000.0
H2CCCCH2 + OH = H2CCCCH + H2O	2.00E+07	2.0	2000.0
H2CCCCH2 + H = H2CCCCH + H2	3.00E+07	2.0	6000.0
CH2CHCCH + OH = HCCHCCH + H2O	7.50E+06	2.0	5000.0
CH2CHCCH + H = HCCHCCH + H2	2.00E+07	2.0	15000.0
CH2CHCCH + OH = H2CCCCH + H2O	1.00E+07	2.0	2000.0
CH2CHCCH + H = H2CCCCH + H2	3.00E+07	2.0	5000.0
HCCHCCH + H = H2CCCCH + H	1.00E+14	0.0	0.0
HCCHCCH + O2 = HCCCHO + HCO	3.00E+12	0.0	0.0
H2CCCCH + O2 = CH2CO + HCCO	1.00E+12	0.0	0.0

Reaction	A	n	E
H2CCCCH + OH = C4H2 + H2O	3.00E+13	0.0	0.0
H2CCCCH + O = CH2CO + C2H	2.00E+13	0.0	0.0
H2CCCCH + O = H2C4O + H	2.00E+13	0.0	0.0
H2CCCCH + H = C4H2 + H2	5.00E+13	0.0	0.0
H2CCCCH + CH2 = aC3H4 + C2H	2.00E+13	0.0	0.0
H2CCCCH (+M) = C4H2 + H (+M)	1.00e14	0.0	47000.
HCCHCCH (+M) = C4H2 + H (+M)	1.00E+14	0.0	36000.0
C4H2 + CH2 = C5H3 + H	1.30E+13	0.0	4326.0
C4H2 + CH = C5H2 + H	1.00E+14	0.0	0.0
C4H2 + CH2SING = C5H3 + H	3.00E+13	0.0	0.0
C4H2 + C2H = C6H2 + H	9.60E+13	0.0	0.0
C4H2 + OH = H2C4O + H	6.66E+12	0.0	-410.0
C4H2 + O = C3H2 + CO	1.20E+12	0.0	0.0
H2C4O + H = C2H2 + HCCO	5.00E+13	0.0	3000.0
H2C4O + OH = CH2CO + HCCO	1.00E+07	2.0	2000.0

IX. APPENDIX B—ADDITIONAL DETAILS CONCERNING THE COFLOW DIFFUSION FLAME EXPERIMENT

A. Mass-Flow Controllers and Their Calibration

The thermal mass-flow controllers (MFCs) were manufactured by MKS Instruments, Inc. Some performance and electrical specifications are listed in Table 19. Each MFC was calibrated for nitrogen by the manufacturer over its designed flow range. Each MFC was also calibrated at NIST/BFRL for the specific gases that it would control.

NIST/BFRL calibrations were performed using an electronic soap-film flow meter, the Gilibrator-2. Using two LED/detectors, the instrument measures the transit time of a soap-film bubble of prescribed diameter over a prescribed distance. The instrument displays the volume flow rate based on the most recent bubble as well as an average over the previous n bubbles. An average over at least 10 bubbles was used for the current calibrations.

Calibration curves were determined for the MFC set point dc voltage (0 V – 5 V) as a function of the volume flow rate. The units of volume flow rate required for input into the calibration curves are standard liters per minute (SLM), Q^*, with a standard condition of 101.3 kPa and 273.15 K. Equation 40 was used to convert the actual desired volume flow rate, Q, to SLM,

$$Q^* = Q \frac{273.15}{T_{amb}} \frac{P_{amb}}{101325}. \quad (40)$$

The units for the ambient temperature (T_{amb}) and pressure (P_{amb}) are Kelvins and Pascals, respectively. Calibration curves were determined for each MFC and for each gas that was controlled by a given MFC. The general calibration curve, described by the following equation,

$$V = A(Q^*)^4 + B(Q^*)^3 + C(Q^*)^2 + DQ^* + E, \quad (41)$$

is a fourth-order polynomial that is used to compute a set point voltage, V, for a desired standard volume flow rate. The coefficients for each MFC and its associated gas(es) are listed in Table 20 to Table 22.

The relative expanded uncertainty of volume flow rate measurements using the Gilibrator-2 has been estimated by Mulholland and Fernandez [171] to be ±1.8 % at a confidence level of 95 % (i.e. a relative combined standard uncertainty of 0.9 %).

B. Detailed Description of Gas Flow Calculations

The gases were assumed to be ideal at constant pressure and temperature and therefore the partial volume flow rates were assumed to sum to the total volume flow rate as described by the Amagat Law. [172] Matching exit velocity conditions required the volume flow rate of the agent/air coflow mixture ($Q_{exit,coflow}$) to remain constant, therefore the flow rate of air was decreased as the flow rate of agent increased. Equations 42 – 44 demonstrate the basis of the software calculations for combining the air (Q_{air}) and agent (Q_{agent}) volume flow rates in the proper proportions to achieve the desired agent volume fraction (X_{agent}) with the required nominal coflow velocity,

Table 19. Mass-Flow Controller Specifications

Specifications	Model 1359C	Model 1559A
Full Scale Range	0-2, 0-10 (SLM, nitrogen)	0-30 (SLM, nitrogen)
Accuracy	± 0.5 % of full scale	± 1.0 % of full scale
Repeatability	± 0.2 % of reading	± 0.2 % of reading
Resolution	± 0.1 % of full scale	± 0.1 % of full scale
Settling Time	< 2 s	< 2 s
Input/Command Signal	0 V - 5 V	0 V - 5 V
Output Signal	0 V - 5 V	0 V - 5 V

Table 20. Calibration Coefficients for the Air and Fuel Mass-Flow Controllers

Coefficient	MFC: Model 1559A, 0 SLM - 30 SLM Gas: Air	MFC: Model 1359C, 0 SLM - 2 SLM Gas: Methane	MFC: Model 1359C, 0 SLM - 2 SLM Gas: Propane
A	1.0937e-6	0.10518	5.8104
B	-6.7825e-5	-0.47988	-8.6052
C	0.0016515	0.62864	3.9647
D	0.15187	3.0543	6.9005
E	0.00099804	0.00093596	0.0010185

Table 21. Calibration Coefficients for Path 1 Agent Mass-Flow Controller

	MFC: Model 1359C, 0 SLM - 10 SLM, Path 1			
	Gas			
Coefficient	Carbon Dioxide	Nitrogen	Helium	Argon
A	-0.0070237	-0.00010244	4.8531e-6	-4.3803e-5
B	0.0091754	0.0022735	-6.8489e-5	0.00086595
C	-0.031546	-0.014304	-0.0041908	-0.00309
D	0.72343	0.52598	0.36329	0.36215
E	-0.0024164	-0.0011183	0.0017435	-0.00026244

Table 22. Calibration Coefficients for Path 2 Agent Mass-Flow Controller

	MFC: Model 1359C, 0 SLM - 10 SLM, Path 2			
	Gas			
Coefficient	Carbon Dioxide	Nitrogen	Helium	Argon
A	-0.00038376	-5.8406e-5	9.8721e-6	-2.656e-5
B	0.0045127	0.0010618	-0.00033397	0.00044042
C	-0.0097097	-0.005722	0.0011492	8.3247e-5
D	0.67999	0.50352	0.33335	0.34989
E	-0.0044515	-0.0040498	0.0031573	-0.0010613

$$Q_{air} + Q_{agent} = Q_{exit,coflow} \qquad (42)$$

$$Q_{agent} = X_{agent} Q_{exit,coflow} \qquad (43)$$

$$Q_{air} = (1 - X_{agent}) Q_{exit,coflow} \qquad (44)$$

Two mass-flow controllers were employed for the agent flow, as displayed in Figure 22. Therefore the Q_{agent} was halved along two paths upstream of the MFC's, Path 1 and Path 2, and then recombined downstream. The input for each agent MFC calibration curve was therefore one half of Q_{agent}.

$$Q_{agent,path1} = Q_{agent,path2} = \frac{Q_{agent}}{2} \qquad (45)$$

The ambient temperature and pressure were recorded for each test and were input into the control program as parameters. The following is an example calculation of the software algorithm used for controlling the fuel and oxidizer flows. The exit velocity for both the fuel and coflow is held constant at 7.0 cm/s, while the coflow is diluted with nitrogen in volume fraction steps of 0.001 until the methane flame is extinguished.

<u>Experiment Constant Conditions</u>

Fuel: Methane (CH_4)
Oxidizer: Air
Agent: Nitrogen
U_{exit}:   7.0 cm/s
$X_{agent,o}$:   0.195
T_{amb}:   295.4 K (22.2 °C)
P_{amb}:   100.0 kPa

<u>User Calculations</u>

Actual volume flow rates:

$$Q_{exit,fuel} = 0.429 \ \frac{L}{min}$$

$$Q_{exit,coflow} = 33.52 \ \frac{L}{min}$$

Standard volume flow rates:

$$Q^*_{fuel} = 0.429 \left(\frac{273.15}{295.4}\right)\left(\frac{100000}{101325}\right) = 0.392 \ \text{SLM}$$

$$Q^*_{coflow} = 33.52 \left(\frac{273.15}{295.4}\right)\left(\frac{100000}{101325}\right) = 30.6 \ \text{SLM}$$

Software Calculations:

Begin iterations
For i = 0 to 30

Fuel MFC set point voltage
$$V_{fuel} = 0.10513 \times (0.392)^4 - 0.47988 \times (0.392)^3$$
$$+ 0.62864 \times (0.392)^2 + 3.0543 \times (0.392) + 0.00093596 = 1.268 \text{ V}$$

Agent volume flow rate:
$$Q^*_{agent} = X_{agent,i} \times Q^*_{coflow} = 0.195 \times 30.6 = 5.97 \text{ SLM}$$
$$Q^*_{agent,path1} = Q^*_{agent,path2} = \frac{5.97}{2} = 2.984 \text{ SLM}$$

Path 1 and 2 Agent MFC set point voltage
$$V_{agent,path1} = -0.00010244 \times (2.984)^4 + 0.0022735 \times (2.984)^3$$
$$- 0.014304 \times (2.984)^2 + 0.52598 \times 2.984 - 0.001183 = 1.493 \text{ V}$$

$$V_{agent,path2} = -5.3406e-5 \times (2.984)^4 + 0.0010618 \times (2.984)^3$$
$$- 0.005722 \times (2.984)^2 + 0.50352 \times 2.984 - 0.0040498 = 1.471 \text{ V}$$

Air volume flow rate:
$$Q^*_{air} = (1 - X_{agent,i}) \times Q^*_{coflow} = (1 - 0.195) \times 30.607 = 24.64 \text{ SLM}$$

Air MFC set point voltage:
$$V_{air} = 1.0937e-6 \times (24.64)^4 - 6.7825e-5 \times (24.64)^3$$
$$+ 0.0016515 \times (24.64)^2 + 0.15187 \times 24.64 + 0.00099804 = 4.134 \text{ V}$$

Elapsed time of 40 seconds on software counter:

Step increase of agent volume fraction:

$$X_{agent,i} = X_{agent,o} + 0.001 = 0.195 + 0.001 = 0.196$$

Return to Beginning of Software Calculations:
Next i

The algorithm continues to increment the agent volume fraction until the flame is extinguished, at which point the operator records the agent concentration and stops the program, or until 30 iterations has been executed and the program automatically shuts off.

C. Uncertainty Analysis

Goals were initially established to meter the flows sufficiently accurately to determine the agent volume fraction in air with an absolute uncertainty of ±0.01. Selection of mass-flow controllers having sufficient accuracy to meet this goal was therefore important. Volumes fractions were calculating using

$$X_{agent} = \frac{Q_{agent,path1} + Q_{agent,path2}}{Q_{air} + Q_{agent,path1} + Q_{agent,path2}}. \quad (46)$$

An uncertainty propagation analysis [173,174,175] was performed on the equation used to calculate the agent concentration. Equation 47 is a generalized form of the error propagation, where R is the calculated result that is a function of the independent variables $z_1, z_2, ..., z_n$ and the uncertainties of those independent variables are $w_1, w_2, ..., w_3$, with w_R being the uncertainty of the calculated result based on the uncertainties in the primary measurements.

$$w_R = \left[\left(\frac{\partial R}{\partial z_1} w_1\right)^2 + \left(\frac{\partial R}{\partial z_2} w_2\right)^2 + ... + \left(\frac{\partial R}{\partial z_n} w_n\right)^2\right]^{1/2} \quad (47)$$

Therefore, taking X_{agent} as the calculated result based on the independently measured volume flow rates, each with an associated uncertainty, the following equations are employed to estimate the uncertainty of the agent concentration. The result, Eq. 48, is the estimated uncertainty for the calculated agent concentration,

$$\frac{\partial X_{agent}}{\partial Q_{agent,path1}} = \frac{\partial X_{agent}}{\partial Q_{agent,path2}} = \frac{Q_{air}}{\left(Q_{air} + Q_{agent,path1} + Q_{agent,path2}\right)^2}$$

$$\frac{\partial X_{agent}}{\partial Q_{air}} = \frac{-\left(Q_{agent,path1} + Q_{agent,path2}\right)}{\left(Q_{air} + Q_{agent,path1} + Q_{agent,path2}\right)^2}$$

$$w_{X_{agent}} = \left[\frac{Q_{air}^2 \left(w_{Q_{agent,path1}}^2 + w_{Q_{agent,path2}}^2\right) + \left(Q_{agent,path1} + Q_{agent,path2}\right)^2 w_{Q_{air}}^2}{\left(Q_{air} + Q_{agent,path1} + Q_{agent,path2}\right)^4}\right]^{1/2}. \quad (48)$$

The MFC's were calibrated within an uncertainty of ± 0.9 % (± 1 σ) for the volume flow rate measurement. This uncertainty is less than those reported by the manufacturer for the MFC. Therefore the standard uncertainty for MFC volume flow rate based on the NIST/BFRL calibration was applied in the uncertainty propagation to compute the agent concentration measurement uncertainty, $w_{X_{agent}}$.

Agent extinguishing concentrations were measured by recording the agent volume fraction at the moment the flame extinguished or at the onset of the lifting instability. Multiple measurements of the agent extinguishing concentration were made in order to provide a significant statistical sample. The measurements were conducted over multiple days and in two separate NIST/BFRL laboratories. The measured standard deviations included in Table 23 for the measurements, $s_{X_{agent,ext}}$, capture the effects of a variable environment and contribute to the overall uncertainty of the agent extinguishing concentration

Table 23. Agent Extinguishing Concentration Measurement Standard Deviation

Agent	Methane		Propane	
	Osc	Ext	Osc	Ext
Carbon Dioxide	0.001	-	0.004	0.001
Nitrogen	0.002	0.002	0.005	0.008
Helium	0.001	0.001	0.003	0.002
Argon	0.001	-	0.002	0.003

Number of samples ≥ 6

Table 24. Uncertainty Budget – Agent Extinguishing Volume Fraction

Source of Uncertainty	Basis	Standard Uncertainty	Type A / B Evaluation [173]
$w_{X_{agent}}$	Uncertainty Propagation (Eq. 48)	±0.002	A
$s_{X_{agent,ext}}$	Statistical	Table 23 (e.g.= 0.002)	A
Combined standard uncertainty: $u_c(X_{agent,ext}) = \sqrt{w_{X_{agent}}^2 + s_{X_{agent,ext}}^2} = \sqrt{0.002^2 + s_{X_{agent,ext}}^2}$			

measurement. Therefore, both the agent concentration measurement uncertainty and the agent extinguishing concentration measurement standard deviation are combined to give the resultant combined standard uncertainties, u_c, for the agent extinguishing volume fraction measurements. The overall uncertainty budget is summarized in Table 24.

In order to obtain a confidence level of 95 %, the combined expanded uncertainty, U, is given by $U = 2u_c$. Values of U have been calculated for the base flame measurements and are included in Table 12.

X. REFERENCES

1. Pitts, W. M., Yang, J. C., Huber, M. L., and Blevins, L. G., *Characterization and Identification of Super-Effective Thermal Fire Extinguishing Agents – First Annual Report*, NISTIR 6414, National Institute of Standards and Technology, Gaithersburg, MD, (October, 1999).
2. Pitts, W. M. and Blevins, L. G., "An Investigation of Extinguishment by Thermal Agents Using Detailed Chemical Kinetic Modeling of Opposed Jet Diffusion Flames," pp. 215-249, Fifteenth Meeting of the UJNR Panel on Fire Research and Safety, NISTIR 6588, National Institute of Standards and Technology, Gaithersburg, MD (November 2000).
3. Pitts, W. M. and Blevins, L. G., "An Investigation of Extinguishment by Thermal Agents Using Detailed Chemical Modeling of Opposed Flow Diffusion Flames," Proceedings of the Halon Options Technical Working Conference 1999, pp. 145-156, Albuquerque, NM, April 27-29, 1999.
4. Pitts, W. M. and Blevins, L. G., "An Investigation of Extinguishment by Thermal Agents Using Detailed Chemical Kinetic Modeling of Opposed Flow Diffusion Flames," pp. 184-187, 1999 Fall Technical Meeting of the Eastern Section of the Combustion Institute, Raleigh, NC, October 11-13, 1999.
5. Yang, J. C., Bryant, R. A., Huber, M. L., and Pitts, W. M., "Experimental Investigation of Extinguishment of Laminar Diffusion Flames by Thermal Agents," pp.433-446, Proceedings of the Halon Options Technical Working Conference 2000, Albuquerque, NM, May 2-4, 2000.
6. Pitts, W. M. and Blevins, L. G., "An Investigation of Extinguishment by Thermal Agents Using Detailed Kinetic Modeling of Opposed Jet Diffusion Flames," Work in Progress Poster 1-K10, Presented at the Twenty-Eight Symposium (International) on Combustion, Edinburgh, Scotland, July 31-August 4, 2000.
7. Pitts, W. M., Bryant, R. A., and Yang, J. C., "Thermal Agent Extinguishment of Two Types of Diffusion Flames," Second Joint Meeting of the United States Sections: The Combustion Institute, March 26-28, 2001, to appear.
8. Pitts, W. M., Yang, J. C., and Bryant, R. A., "Fuel Effects on the Extinguishment of Laminar Diffusion Flames by Thermal Agents," Proceedings of the Halon Options Technical Working Conference 2001, Albuquerque, NM, April 24-26, 2001, to appear.
9. *DIPPR Data Compilation of Pure Compound Properties Database*, V9.02, NIST Standard Reference Data Program #11, NIST, Gaithersburg MD (1995).
10. *NIST Thermodynamic Properties of Refrigerants and Refrigerant Mixtures Database (REFPROP)*, Version 5.0, NIST Standard Reference Data Program #23, NIST, Gaithersburg MD (1996), Version 6.0, (1998).
11. Heathfield, A. E., Anastasi, C., Pagsberg, P., and McCulloch, A., *Atmosph. Envir.* **32**, 711-717 1998.
12. Grzyll, L. R., Back, D. D., Ramos, C., and Samad, N.A., *Development of a Rapid Screening Technique for Second-Generation Halon Alternatives*, SBIR Phase II Final Report, Contract Number DAAH01-93-C-R150, (1996).
13. Minor, B., *Compositions Including a Three Carbon Cyclic Fluoroether*, U.S. Patent No. 5480572 (1996).
14. Minor, B., *Compositions Including a Fluoroamine and a Second Component*, U.S. Patent No. 5441659 (1995).
15. Minor, B., Chisolm, T. E. C., and Shealy, G. S., *Azeotrope (Like) Compositions with Fluoromethyl Trifluoromethyl Ether and Dimethyl Ether*, U.S. Patent No, 5607616 (1997).
16. Pitts, W. M., Nyden, M. R., Gann, R. G., Mallard, W. G. and Tsang, W., *Construction of an Exploratory List of Chemicals to Initiate the Search for Halon Alternatives*, NIST Technical Note 1279 (1990).
17. Sekiya, A. and Misaki, S., *Chemtech* **26**, 44-48 (1996).

18. NIST *Structures and Properties Database and Estimation Program*, NIST Standard Reference Data Program #25, Version 2.02, National Institute of Standards and Technology, Gaithersburg, MD (1994).
19. Fishtine, S. H., *Ind. Eng. Chem.* **55**, 20-28 (1963).
20. Hawley, G. G., *Condensed Chemical Dictionary, Eighth Ed.*, Van Nostrand Reinhold Company, New York (1971).
21. Chueh, C. F. and Swanson, A. C., *Chem. Eng. Progress* **69**, 83-85 (1973); *Can. J. Chem. Eng.* **51**, 596-600 (1973).
22. *NIST Chemistry Webbook*, http://webbook.nist.gov.
23. 3M Corporation; web site http://www.mmm.com/fluids/7100.html.
24. Salvi-Narkhede, M., Adcock, J. L., Gakh, A., and Van Hook, W. A., *J. Chem. Thermodynamics* **25**, 643-647 (1993).
25. Wang, B. H., Adcock, J. L., Mathur, S. B., and Van Hook, W. A., *J. Chem. Thermodynamics* **3**, 699-710 (1991).
26. Matsuo, S., Tanaka, Y., Takada, N., Yamamoto, H., and Sekiya, A., *J. Chem. Eng. Data* **43**, 473-476 (1998).
27. Defibaugh, D. R., Carillo-Nava, E., Hurly, J. J., Moldover, M. R., Schmidt, J. W., and Weber, L. A., *J. Chem. Eng. Data* **42**, 488-496 (1997).
28. *Aldrich Fluorinated Products Catalog*, 1998-1999, Aldrich Chemical Company, Milwaukee, WI.
29. Defibaugh, D. R., Gillis, K. A., Moldover, M. R., Morrison, G., and Schmidt, J. W., *Fluid Phase Equilibria* **81**, 285-305 (1992).
30. Rihani, D. N. and Doraiswamy, L. K., *Ind. Eng. Chem. Fund.* **4,** 17-21 (1965).
31. Apollo Scientific, Derbyshire, UK; http://www.apolloscientific.co.uk/.
32. Flynn, R. M. and Thomas, S. D., *Fire Extinguishing Process and Composition*, U. S. Patent 5,718,293, Feb ruary 17, 1998.
33. Hamins, A., Gmurczyk, G., Grosshandler, W., Rehwoldt, R. G., Vazquez, I., Cleary, T., Presser, C., and Seshadri, K., "4. Flame Suppression Effectiveness," in *Evaluation of Alternative In-Flight Fire Suppressants for Full-Scale Testing in Simulated Aircraft Engine Nacelles and Dry Bays* (Grosshandler, W. L., Gann, R. G., and Pitts, W. M., Eds.), NIST SP 861, pp. 345-465, National Institute of Standards and Technology, Gaithersburg, MD (April, 1994).
34. Abderafi, S. and Bounahmidi, T., *Fluid Phase Equilibria* **162**, 225-240 (1999).
35. Yang, J. C., Donnelly, M. K., Privé, N. C., and Grosshandler, W. L., *Fire Safety J.* **36**, 55-72 (2001).
36. Yang, J. C., Donnelly, M. K., Privé, N. C., and Grosshandler, W. L., *Dispersed Liquid Agent Fire Suppression Screen Apparatus*, NISTIR 6319, National Institute of Standards and Technology, Gaithersburg, MD (July, 1999).
37. Grosshandler, W. L., Selepak, M., Donnelly, M., Charagundla, R., and Presser, C., "Suppressant Performance Evaluation in a Baffle-Stabilized Pool Fire," pp. 105-116, *Halon Options Technical Working Conference*, Albuquerque, NM (April 27-29, 1999).
38. Grosshandler, W. L., Hamins, A., Charagundla, R., and Presser, C., "Suppression Effectiveness Screening for Impulsively Discharged Agents," pp. 15-25, *Halon Options Technical Working Conference*, Albuquerque, NM (May 2-4, 2000).
39. Grosshandler, W. L., Hamins, A., McGrattan, K. B., and Presser, C., *Transient Application, Recirculating Pool Fire Agent Effectiveness Screen: Final Report. NGP Project 3A/2/890*, NISTIR 6733, National Institute of Standards and Technology, Gaithersburg, MD, (April, 2001).
40. White, A. G., *J. Chem. Soc.* **127**, 672-684 (1925).
41. Simmons, R. F. and Wolfhard, H. G., *Combust. Flame* **1**, 155-161 (1957).
42. Maĉek, A., *Combust. Sci. Tech.* **21**, 43-52 (1979).
43. Huggett, C., *Combust. Flame* **20**, 140-142 (1973).
44. McHale, E. T. *Fire Technol.* **10**, 15-24 (1974).
45. Tucker, D. M., Drysdale, D. D., and Rasbash, D. J., *Combust. Flame* **41**, 293-300 (1981).

46. Bajpai, S. N., *J. Fire Flamm.* **5**, 255-267 (1974).
47. Hirst, R. and Booth, K., *Fire Technol.* **13**, 296-315 (1977).
48. Ishizuka, S. and Tsuji, H., *Eighteenth Symp. (Intl.) Combust.*, pp. 695-703, The Combustion Institute, Pittsburgh (1981).
49. Sheinson, R. S., Penner-Hahn, J. E., and Indritz, D., *Fire Safety J.* **15**, 437-450 (1989).
50. Roberts, A. F. and Quince, B. W., *Combust. Flame* **20**, 245-251 (1973).
51. Dodding, R. A., Simmons, R. F., and Stephens, A., *Combust. Flame* **15**, 313-315 (1970).
52. Zegers, E. J. P., Williams, B. A., Fisher, E. M., Fleming, J. W., and Sheinson, R. S., *Combust. Flame* **121**, 471-487 (2000).
53. Hamins, A., Trees, D., Seshadri, K., and Chelliah, H. K, *Combust. Flame* **99**, 221-230 (1994).
54. Moore, T. A., Weitz, C. A., and Tapscott, R. E., "An Update of NMERI Cup-Burner Test Results," pp. 551-564, *Halon Options Technical Working Conference*, Albuquerque, NM (May 7-9, 1996).
55. Saito, N., Ogawa, Y., Saso, Y., Liao, C., and Sakei, R., *Fire Safety J.* **27**, 185-200 (1996).
56. Babb, M., Gollahalli, S. R., and Sliepcevich, C. M., *J. Propulsion Power* **15**, 260-265 (1999).
57. Ural, E. A., "Measurements of the Extinguishing Concentrations of Gaseous Fuels Using the Cup Burner Apparatus," pp. 275-283, *Halon Options Technical Working Conference*, Albuquerque, NM (April 27-29, 1999).
58. Larsen, E. R., *J. Fire Flamm./Fire Retard. Chem.* **2**, 5-20 (1975).
59. Larsen, E. R., "Halogenated Fire Extinguishants: Flame Suppression by a Physical Mechanism?," in *Halogenated Fire Suppressants* (Gann, R. G., Editor), pp. 376-388, American Chemical Society, Washington, DC (1975).
60. Ewing, C. T., Hughes, J. T., and Carhart, H. W., *Fire Materials* **8**, 148-156 (1984).
61. Ewing, C. T., Beyler, C. L., and Carhart, H. W., *J. Fire Prot. Eng.* **6**, 23-54 (1994).
62. Westbrook, C. K., *Combust. Science Tech.* **23**, 191-202 (1980).
63. Westbrook, C. K., *Nineteenth Symp. (Intl.) Combust.*, pp. 127-141, The Combustion Institute, Pittsburgh (1982).
64. Westbrook, C. K., *Combust. Science Tech.* **34**, 201-225 (1983).
65. Noto, T., Babushok, V. I., Burgess, D. R. F., Hamins, A., Tsang, W., and Miziolek, A., *Twenty-Sixth Symp. (Intl.) Combust.*, pp. 1377-1383, The Combustion Institute, Pittsburgh (1996).
66. Noto, T., Babushok, V., Hamins, A., and Tsang, W., *Combust. Flame* **112**, 147-160 (1998).
67. Linteris, G. T., Burgess, Jr., D. R., Babushok, V., Zachariah, M., Tsang, W., and Westmoreland, P., *Combust. Flame* **113**, 164-180 (1998).
68. Saso, Y., Zhu, D. L., Wang, H., Law, C. K., and Saito, N., *Combust. Flame* **114**, 457-468 (1998).
69. Rumminger, M. D., Reinelt, D., Babushok, V., and Linteris, G. T., *Combust. Flame* **116**, 207-219 (1999),
70. Babushok, V. and Tsang, W., *Combust. Flame* **123**, 488-506 (2000).
71. Hamins, A., Trees, D., Seshadri, K., and Chelliah, H. K., *Combust. Flame* **99**, 231-239 (1994).
72. Tanoff, M. A., Dobbins, R. R., Smooke, M. D., Burgess, Jr., D. R., Zachariah, M. R., and Tsang, W., "C_1 and C_2 Fluorinated Hydrocarbon Effects on the Extinction Characteristics of Methane Vs. Air Counterflow Diffusion Flames," pp. 116-126, *Halon Options Technical Working Conference*, Albuquerque, NM (May 6-8, 1997).
73. Lentati, A. M. and Chelliah, H. K., *Combust. Flame* **115**, 158-179 (1998).
74. Lentati, A. M. and Chelliah, H. K., *Twenty-Seventh Symp. (Intl.) Combust.*, pp. 2839-2846, The Combustion Institute, Pittsburgh (1998).
75. Tuovinen, H., *The Computer Modelling of the Combustion and Extinction of Methane-Air Premixed Flame, By Thermally Stable Gase*, SP Report 1989:54, Swedish National Testing Institution, Borås, Sweden (1989).
76. Peters, N., *Prog. Energy Combust. Sci.* **10**, 319-339 (1984).
77. Peters, N, *Twenty-First Symp. (Intl.) Combust.*, pp. 1231-1250, The Combustion Institute, Pittsburgh, (1986).

78. Tsuji, H., *Prog. Energy Combust. Sci.* **8**, 93-119 (1982).
79. Dixon-Lewis, G., *Twenty-Third Symp. (Intl.) Combust.*, pp. 305-324, The Combustion Institute, Pittsburgh (1990).
80. Potter, Jr., A. E., Heimel, S., and Butler, J. N., *Eighth Symp. (Intl.) Combust.*, pp. 1027-1034, Williams and Wilkins, Baltimore (1962).
81. Friedman, R. and Levy, J. B., *Combust. Flame* **7**, 195-201 (1963).
82. Ibiricu, M. M. and Gaydon, A. G., *Combust. Flame* **19**, 51-62 (1964).
83. Dodding, R. A., Simmons, R. F., and Stephens, A., *Combust. Flame* **15**, 313-315 (1970).
84. Milne, T. A., Green, C. L., and Benson, D. K., *Combust. Flame* **15**, 255-263 (1970).
85. Papas, P., Fleming, J. W., and Sheinson, R. S., *Twenty-Sixth Symp. (Intl.) Combust.*, pp. 1405-1411, The Combustion Institute, Pittsburgh (1996).
86. MacDonald, M. A., Jayaweera, T. M, Fisher, E. M., and Gouldin, F. C., *Combust. Flame* **116**, 166-176 (1999).
87. Hahn, W. A. and Wendt, J. O. L., *Eighteenth Symp. (Intl.) Combust.*, pp. 121-131, The Combustion Institute, Pittsburgh (1981).
88. Dixon-Lewis, G., David, T., Gaskell, P. H., Fukutani, S., Jinno, H., Miller, J. A., Kee, R. J., Smooke, M. D., Peters, N., Effelsberg, E., Warnatz, J., and Behrendt, F., *Twentieth Symp. (Intl.) Combust.*, pp. 1893-1904, The Combustion Institute, Pittsburgh (1984).
89. Smooke, M. D., Puri, I. K., and Seshadri, K., *Twenty-First Symp. (Intl.) Combust.*, pp. 1783-1792, The Combustion Institute, Pittsburgh, (1986).
90. Dixon-Lewis, G., David, T., and Gaskell, P. H., *Archivum Combustionis* **6**, 3 - 21 (1986).
91. Puri, I. K., Seshadri, K., Smooke, M. D., and Keyes, D. E., *Combust. Sci. Tech.* **56**, 1-22 (1987).
92. Chelliah, H. K., Law, C. K., Ueda, T., Smooke, M. D., and Williams, F. A., *Twenty-Third Symp. (Intl.) Combust.*, pp. 503-511, The Combustion Institute, Pittsburgh (1990).
93. Law, C. K., *Twenty-Second Symp. (Intl.) Combustion*, pp. 1381-1402, The Combustion Institute, Pittsburgh (1988)
94. Tsuji, H. and Yamaoka, I. *Twelfth Symp. (Intl.) Combust.*, pp. 997-1005, The Combustion Institute, Pittsburgh (1969).
95. Tsuji, H. and Yamaoka, I., *Thirteenth Symp. (Intl.) Combust.*, pp. 723-731, The Combustion Institute, Pittsburgh (1971).
96. Kee, R. J., Miller, J. A., Evans, G. H., and Dixon-Lewis, G., *Twenty-Second Symp. (Intl.) Combust.*, pp. 1479-1494, The Combustion Institute, Pittsburgh (1988).
97. Seshadri, K. and Williams, F. A., *Int. J. Heat Mass Transfer* **21**, 251-253 (1978).
98. Decroix, M. E. and Roberts, W. L., *Combust. Science Tech.* **146**, 57-84 (1999).
99. Fisher, E. M., Williams, B. A., and Fleming, J. W., "Determination of the Strain in Counterflow Diffusion Flames From Flow Conditions," *1997 Fall Technical Meeting of the Eastern States Section of the Combustion Institute*, pp. 191-194, Hartford, CT (Oct. 27-29, 1997).
100. Maruta, K., Yoshida, M., Guo, H., Ju, Y., and Niioka, T., *Combust. Flame* **112**, 181-187 (1998).
101. Chan, S. H., Yin, J. Q., and Shi, B. J., *Combust. Flame* **112**, 445-456 (1998).
102. Gore, J. P., Lim, J., Takeno, T., and Zhu, X. L., Paper AJTE99-6311, Proceedings of the 5th ASME/JSME Joint Thermal Engineering Conference, San Diego, CA (March 15-19, 1999).
103. Bai, X. S., Fuchs, L., and Mauss, F., *Combust. Flame* **120**, 285-300 (1999).
104. Lutz, A. E., Kee, R. J., Grcar, J. F., and Rupley, F. M., *OPPDIF: A Fortran Program for Computing Opposed-Flow Diffusion Flames*, SAND96-8243, Sandia National Laboratories, Livermore, CA (1996).
105. Kee, R. J., Rupley, F. M., Meeks, E., and Miller, J. A., *CHEMKIN-III: A FORTRAN Chemical Kinetics Package for the Analysis of Gas-Phase Chemical and Plasma Kinetics*, SAND96-8216, Sandia National Laboratories, Livermore, CA (1996).

106. Frenklach, M., Wang, H., Yu, C.-L., Goldenberg, M., Bowman, C. T., Hanson, R. K., Davidson, D. F., Chang, E. J., Smith, G. P., Golden, D. M., Gardiner, W. C., and Lissianski, V., *GRI-MECH 1.2*, http://www.me.berkeley.edu/gri_mech/
107. Marinov, N. M., Pitz, W. J., Westbrook, C. K., Castaldi, M. J., and Senkan, S. M., *Combust. Sci. Technol.* **116,117**, 211-287 (1996).
108. Marinov, N. M., Pitz, W. J., Westbrook, C. K., Vincitore, A. M., Castaldi, M. J., and Senkan, S. M., *Combust. Flame* **114**, 192-213 (1998).
109. Fendell, F. E., *J. Fluid Mech.* **21**, 281-303 (1965).
110. Liñán, A., *Acta Astronautica* **1**, 1007-1039 (1974).
111. Giovangigli, V. and Smooke, M. D., *Combust. Science Tech.* **53**, 23-49 (1987).
112. Nishioka, M. and Law, C. K., *Combust. Flame* **104**, 328-342 (1996).
113. Norton, T. S., Smyth, K. C., Miller, J. H., and Smooke, M. D., *Combust. Science Technol.* **90**, 1-34 (1993).
114. Puri, I. K. and Seshadri, K., *Combust. Flame* **65**, 137-150 (1986).
115. Du. D. X., Axelbaum, R. L., and Law, C. K., *Twenty-Second Symp. (Intl.) Combust.*, pp. 387-394, The Combustion Institute, Pittsburgh (1988).
116. Yang, G. and Kennedy, I. M., *Combust. Flame* **92**, 187-196 (1993).
117. Du, J. and Axelbaum, R. L., *Twenty-Sixth Symp. (Intl.) Combust.*, pp. 1137-1142, The Combustion Institute, Pittsburgh (1996).
118. Pellett, G. L., Isaac, K. M., Humphreys, Jr., W. M., Gartrell, L. R., Roberts, W. L., Dancy, C. L., and Northam, G. B., *Combust. Flame* **112**, 575-592 (1998).
119. Peters, N., "Flame Calculations with Reduced Mechanisms," in *Reduced Mechanisms for Applications in Combustion Systems*, (Peters, N. and Rogg, B., eds.), pp. 1-14, Springer, New York (1993).
120. Lin, K.-C. and Faeth, G. M., *J. Propul. Power* **12**, 691-698 (1996).
121. A. Hamins, Building and Fire Research Laboratory, National Institute of Standards and Technology, Gaithersburg, MD, private communication.
122. Santoro, R. J., Semerjian, H. G., and Dobbins, R. A., *Combust. Flame* **51**, 203-218 (1983).
123. Smyth, K. C. and Everest, D. A., *Twenty-Sixth Symp. (Intl.) Combust.*, pp. 1385-1393, The Combustion Institute, Pittsburgh, (1996).
124. Smyth, K. C., Harrington, J. E., Johnsson, E. L., and Pitts, W. M., *Combust. Flame* **95**, 229-239 (1993).
125. Pitts, W. M., Mullholland, G. W., Breuel, B. D., Johnsson, E. L., Chung, S., Harris, R., Hess, D. E., "Real-Time Suppressant Concentration Measurement," in Fire Suppression System Performance of Alternative Agents in Aircraft Engine and Dry Bay Laboratory Simulations, (Gann, R. G., Ed.), National Institute of Standards and Technology SP-890, Vol. II, pp. 319-585, (1995).
126. Chan, W. Y. and T'ien, J. S., *Combust. Sci. Tech.* **18**, 139-143 (1978).
127. Dietrich, D. L., Ross, H. D., and T'ien, J. S., "Candle Flames in Microgravity," in *Proceedings of the Third International Microgravity Workshop*, NASA Conference Publication 10174, pp. 31-36, Cleveland, OH (1995).
128. Chung, S. H. and Lee, B. J., *Combust. Flame* **86**, 62-72 (1991).
129. Lee, B. J. and Chung, S. H., *Combust. Flame* **109**, 163-172 (1997).
130. Lee, B. J., Kim, J. S., and Chung, S. H., *Twenty-Fifth Symp. (Intl.) Combust.*, pp. 1175-1181, The Combustion Institute, Pittsburgh (1994).
131. Brooker J. E., Jia, K., Stocker, D. P., and Chen, L.-D., "Influence of Buoyant Convection on the Stability of Enclosed Laminar Flames," in *Proceedings of the Fifth International Microgravity Combustion Workshop*, NASA Conference Publication 208917, pp. 97-100, Cleveland, OH (1999).
132. Huerre, P. and Monkewitz, P. A., *Annu. Rev. Fluid Mech.* **22**, 473-537 (1990).
133. Liepmann, D. and Gharib, M., *J. Fluid Mech.* **245**, 643-668 (1992).

134. Lasheras, J. C., Liñán, A., Lecuona, A., and Rodriguez, P., *Twenty-Fourth Symp. (Intl.) Combust.*, pp. 325-332, The Combustion Institute, Pittsburgh, (1992).
135. Saso, Y., Saito, N., and Iwata, Y., *Fire Tech.* **29**, 22-33 (1993).
136. Saso, Y., Saito, N., Liao, C., and Ogawa, Y., *Fire Safety J.* **26**, 303-326 (1996).
137. Coward, H. F. and Hartwell, F. J., *J. Chem. Soc.*, 1522-1532 (1926).
138. Liñán, A. and Williams, F. A., *Fundamental Aspects of Combustion*, Oxford University Press, New York (1993).
139. Gordon, A. S., Smith, S. R., and McNesby, J. R., *Seventh Symp. (Intl.) Combust.*, pp. 317-324, Butterworths, London (1959).
140. Tapscott, R. E., Personal Communication, February 26, 1999, *NMERI In-House Testing*, June 2, 1997.
141. Flynn, R. M. and Thomas, S. D., *Fire Extinguishing Process and Composition*, U. S. Patent 5,718,293 (February 17, 1998).
142. Downie, B., Polymeropoulos, C., and Gogos, G., *Fire Safety J.* **24**, 359-381 (1995).
143. Rasbash, D. J., *Proceedings of the First International Symposium of Fire Safety Science*, pp. 1145-1163, Hemisphere, Washington (1986).
144. Engel, O. G., *J. Res. NBS* **54**, 281-298 (1955).
145. Chandra, S. and Avedisian, C. T., *Proc. R. Soc. Lond.* **A432**, 13-41 (1991).
146. Chandra, S. and Avedisian, C. T., *Int. J. Heat Mass Transfer* **35**, 2377-2388 (1992).
147. Mundo, C., Sommerfeld, M., and Tropea, C., *Int. J. Multiphase Flow* **21**, 151-173 (1995).
148. Yarin, A. L. and Weiss, D. A., *J. Fluid Mech.* **283**, 141-173 (1995).
149. Bernardin, J. D. and Mudawar, I., *Int. J. Heat Mass Transfer* **39**, 2023-2037 (1996).
150. Bernardin, J. D., Stebbins, C. J., and Mudawar, I., *Int. J. Heat Mass Transfer* **40**, 73-88 (1997).
151. Bernardin, J. D., Stebbins, C. J., and Mudawar, I., *Int. J. Heat Mass Transfer* **40**, 247-267 (1997).
152. Bernardin, J. D. and Mudawar, I., *Int. J. Heat Mass Transfer* **40**, 2579-2593 (1997).
153. Mudawar, I. and Valentine, W. S., *J. Heat Treating* **7**, 107-121 (1989).
154. Klinzing, W. P., Rozzi, J. C., and Mudawar, I., *J. Heat Treating* **9**, 91-103 (1992).
155. Mudawar, I. and Deiters, T. A., *Int. J. Heat Mass Transfer* **37**, 347-362 (1994).
156. Hall, D. D. and Mudawar, I., *Int. J. Heat Mass Transfer* **38**, 1201-1216 (1995).
157. Estes, K. A. and Mudawar, I., *Int. J. Heat Mass Transfer* **38**, 2985-2996 (1995).
158. Mudawar, I. and Estes, K. A., *ASME J. Heat Transfer* **118**, 672-679 (1996).
159. Rein, M., *Fluid Dynamic Research* **12**, 61-93 (1993).
160. Henry, R. E. and Fauske, H. K., *ASME J. Heat Transfer* **101**, 280-287 (1979).
161. Spalding, D. B., *Some Fundamentals of Combustion*, Butterworths Scientific Publications, London (1955).
162. Kanury, A. M., *Introduction to Combustion Phenomena*, Gordon & Breach, New York (1975).
163. Chin, J. S. and Lefebvre, A. H., *AIAA J.* **21**, 1437-1443 (1983).
164. Reid, R. C., Prausnitz, J. M., and Poling, B. E., *The Properties of Gases & Liquids*, 4th ed., McGraw-Hill, New York (1987).
165. Lefebvre, A. H., *Atomization and Sprays*, Taylor & Francis, Washington (1989).
166. Qiao, Y. M., and Chandra, S., *ASME J. Heat Transfer* **120**, 92-98 (1998).
167. Deb, S. and Yao, S.-C., *Int. J. Heat Mass Transfer* **32**, 2099-2112 (1989).
168. Seshadri, K., *Combust. Flame* **33**, 197-215 (1978).
169. Zegers, E. J. P., Williams, B. A., Sheinson, R. S., and Fleming, J. W., *Twenty-Eight Symp. (Intl.) Combust.*, pp. 2931-2937, The Combustion Institute, Pittsburgh (2000).
170. Lazzarini, A. K., Krauss, R. H., Chelliah, H. K., and Linteris, G. T., *Twenty-Eight Symp. (Intl.) Combust.*, pp. 2939-2945, The Combustion Institute, Pittsburgh (2000).
171. Mulholland, G. and Fernandez, M., "Report on the Calibration of the Gilibrator-2 Soap Film Flowmeter," NIST/BFRL Internal Memo. (1997).

172. Moran, M. J. and Shapiro, H. N., *Fundamentals of Engineering Thermodynamics*, John Wiley & Sons, Inc., New York, (1988).
173. Taylor, B. N. and Kuyatt, C. E., *Guidelines for Evaluating and Expressing the Uncertainty of NIST Measurement Results*, NIST Technical Note 1297, National Institute of Standards and Technology, Gaithersburg, MD (September, 1994).
174. Taylor, J. R., *An Introduction to Error Analysis – The Study of Uncertainties in Physical Measurements*, 2nd Edition, University Science Books, California, pp. 73-79 (1997).
175. Holman, J. P., *Experimental Methods for Engineers*, 4th Edition, pp. 50-57, McGraw-Hill, New York, (1984).

www.ingramcontent.com/pod-product-compliance
Lightning Source LLC
Chambersburg PA
CBHW081726170526
45167CB00009B/3715